Falsework

A handbook of design and practice

Falsework

A Handbook of Design and Practice

A. W. Irwin
W. I. Sibbald

GRANADA
London Toronto Sydney New York

Granada Technical Books
Granada Publishing Ltd
8 Grafton Street, London W1X 3LA

First published in Great Britain 1983 by
Granada Publishing

British Library Cataloguing in Publication Data

Irwin, A.W.
    Falsework.
    1. Scaffolding
    2. Shoning and underpinning
    I. Title   II. Sibbald, W.I.
    624.1'6   TH5281

ISBN 0-246-11809-1

Printed and bound in Great Britain by Adlard & Son Ltd, Dorking

Contents

Acknowledgements   vii
Preface   ix

1 Materials, Finishes, Special Formwork   1

1.1   Materials   1
1.2   Surface finishes   7
1.3   Custom made and special purpose formwork   8

2 Column and Wall Formwork   23

2.1   Columns   23
    2.1.1   Column formwork components   23
2.2   Wall formwork   26
    2.2.1   Wall formwork components   26
2.3   Proprietary wall formwork components   31
    2.3.1   Proprietary soldiers   31
    2.3.2   Proprietary wall formwork systems   35
2.4   Concrete pressures   37
2.5   Wall formwork calculations – Examples   40
    2.5.1   Example 1   40
    2.5.2   Example 2   41

3 Deck and Floor Falsework   52

3.1   Floor slabs   52
3.2   Formwork materials and components   53
3.3   Modular decking systems   56
3.4   Support and centering materials   56
    3.4.1   Scaffold tube and fittings to BS 1139   56
    3.4.2   Adjustable steel props to BS 4074   58
    3.4.3   Proprietary soffit beams – non-system   59
    3.4.4   Proprietary scaffold support systems   60
    3.4.5   Heavy duty shores and towers   62
    3.4.6   Heavy duty bridging and arch girders   63
3.5   Design example using simplified method for a load-bearing birdcage scaffold   64
    3.5.1   Introduction to method   64

3.5.2	Example	65
3.5.3	Design procedure	65
3.5.4	Design procedure – discussion and explanation	71

4   Falsework Supervision and Procedures   84

5   Bridge Falsework   90

5.1	Design methods	90
5.2	Force actions	90
	5.2.1 Main loadings	91
	5.2.2 Environmental effects	92
	5.2.3 Constraint effects	98
5.3	Alternative falsework schemes for construction of a bridge	98
	5.3.1 General bridge falsework scheme 1 – Centre span	99
	5.3.2 General scheme 2	113
	5.3.3 General scheme 3	113
	5.3.4 General schemes 4a and 4b	117

6   Design Checks and Safety Aspects for Falsework and Access Scaffolding   118

6.1	General	118
6.2	Common problem areas	118
6.3	Simple falsework check procedure	121

7   References   123

Appendix A – Data for Formwork Design   125

Appendix B – Approximate Analysis of Trusses by the Method of Sections   140

Appendix C – Approximate Frame Deflections by the Flexibility Method   142

Appendix D – Simple Analysis of Members for Torsion   143

Appendix E – Wind Force Data   145

Appendix F – General Design Aids and Data Tables   148

Appendix G – Definitions   171

Index   175

Acknowledgements

The authors are indebted to those companies and organisations listed below for their assistance in the preparation of this book and for permission to publish copyright material. Sources of illustrations are gratefully acknowledged.

Acrow (Engineers) Ltd – figs. 2.1, 2.8, 2.9, 3.3, 3.8

Balfour Beattie Construction Ltd

Boskalis Westminster Ltd – Plate 15

British Standards Institution (2 Park Street, London, W1A 2BS from whom complete copies can be obtained) – figs. A.5, E.1 (adapted), F.2; tables 1.1 – 1.4, 3.2, F.2, F.13, F.15 – F.20 and extracts from BS 5975: 1982; table E.1 from BS CP3; Chapter V; Part 2: 1972; tables F.8 – F.10 from BS 449: 1970; table A.4 from BS CP 112: Part 2: 1971; and extracts from the BSI Draft Code of Practice for Falsework, 1975

Cement and Concrete Association – Plates 2, 3, 5 – 10

Concrete Society – table 1.6

Construction Industry Research and Information Association – table 2.2

Council of Forest Industries of British Columbie – table A.3

Expanded Metal Company Ltd – fig. 3.2; Plates 13, 14

Film Board of Canada – Plate 11

Finnish Plywood Development Association – fig. A.4; tables 1.5 and A.2

GKN Mills Ltd – Plates 38, 43, 46

HMSO – fig. 5.3. Crown copyright

Kwikform Ltd – fig. 3.6, Plates 26, 27, 35, 36, 39, 40, 47, 51

Mabey Bridge Company Ltd

North of Scotland Hydro Electric Board – Plates 24, 25, 31, 32

Rapid Metal Developments Ltd – fig. 2.6; Plates 33, 34, 41, 44, 45

Scaffolding (Great Britain) Ltd – table 2.1

Scott Wilson Kirkpatrick and Partners – Plates 16, 42, 48

Stelmo Ltd – Plates 17 – 23, 28 – 30

Preface

It was one of the principal recommendations in the final report of the Advisory Committee on Falsework[20] that 'Short professional course in falsework be provided for practising engineers and architects'. In response the authors organised an in-service course based on the Draft Falsework Code.[17] These course lecture notes, revised following the publication of BS 5975: 1982,[1] and expanded to include more information on falsework materials are the basis of this book.

The target readership is similar to the original course, it is intended primarily to be used as a practical handbook for falsework designers and others with supervisory or checking responsibilities. An additional aim is to present basic design principles and construction methods in formwork and centering for students of civil engineering and building at degree and further education level.

Recommended design methods are illustrated by worked examples and draw on the data in the Appendixes. The provisions in The British Standards Institution's Code of Practice for Falsework, BS 5975: 1982 have been used in the approach to analysis and design, also in the loadings and stresses used. The code should be used in conjunction with this book in the design of falsework.

In order to give the reader a general overview of the types of product available in the industry a selection of proprietary falsework material is described and the principal characteristics given. This list is not comprehensive, reference should be made to the manufacturers for detailed information about their products.

Many developments taking place in the formwork, falsework and scaffolding industry are referred to in this book, for example the increasing use of aluminium and new techniques in formwork lining. If engineers and others concerned are to promote improvements in cost, safety and quality of finish they will have to keep themselves informed about advances in procedure, design methods and new materials.

A.W. Irwin
W.I. Sibbald

Chapter 1
Materials, Finishes, Special Formwork

1.1 MATERIALS

Timber

Until very recently many types of temporary works, including falsework, were described as 'timbering'. Figure 1.1 which illustrates the centering for a masonry bridge, shows clearly the origin of the term. Another interesting point about this 1866 print of Smeaton's bridge is the natural manner in which the permanent and temporary works are considered together by the designer. There is no sign here of the separation of design from construction which was to take place in a less self-confident era of civil engineering.

Rules of thumb were used in early timber centering, to be followed by calculation based on elementary structural principles and traditional stress values. Todays falsework

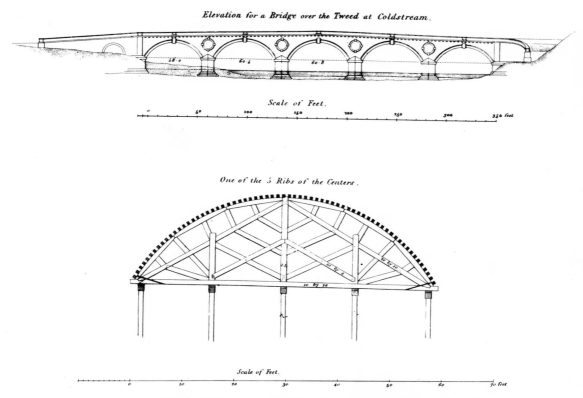

Figure 1.1 Centering for a bridge over the Tweed at Coldstream

designer is encouraged to realise the full potential of timber as a construction material by using stresses matched to quality grading with appropriate modifications to allow for the temporary nature of its use in falsework structures.

The basic strength properties of timber, hardwood or softwood, when used as a structural element are determined by:

(a) Timber species.
(b) Degree of seasoning as measured by moisture content. Dry timber is significantly stronger than green. Falsework timber may be assumed to be in 'wet' condition with a moisture content greater than 18%.
(c) The presence of defects and other physical characteristics which affect strength. These are knots, fissures, slope of grain, density of growth rings, resin pockets, distortion and loss of cross section (wane).

All of these features can be measured for visual stress grading in accordance with BS 4978[21] or any other appropriate standard. Alternatively, timber can be machine stress graded by an approved machine which will measure one or more of the indicating properties in a non-destructive test.

The timber stresses recommended in the Falsework Code are based on BS 5268, 'Code of practice for the structural use of timber', which supersedes BS 112 and classifies timber into eight strength grades irrespective of species. (BS 5268 not published at time of writing.) Although any timber may be used in falsework provided the appropriate stresses are used it is recommended that generally strength class 3 and above should be used (table 1.1).

The species and grade classification used in BS 112[22] converts directly to strength class (table 1.2). Alternatively, timber may be graded directly into a particular strength class.

The working stresses given in BS 5975 follow the trend towards limit state design and are based on a characteristic value of 95%, i.e. 5% of test samples would fail to reach the working stress.

Table 1.1 Grade stresses and moduli of elasticity for the wet condition (reference 1)

Strength class	Bending stress parallel to grain (N/mm²)	Tension stress parallel to grain (N/mm²)	Compression stress perpendicular to grain (N/mm²)	Shear stress parallel to grain (N/mm²)	Modulus of elasticity mean (N/mm²)	minimum (N/mm²)
SC3	4.08	2.48	1.32	0.612	6800	4480
SC4	5.84	3.52	1.44	0.612	7440	5200
SC5	8.00	4.80	1.68	0.801	8720	5680

1. This table is based on the proposed revision of CP 112: Part 2. The values take into account the modification factor for the wet condition in that code.
2. Timber in the 'wet condition' has a moisture content greater than 18%.

The grade stresses given in table 1.1 are for timber assumed to be performing an engineering function in a permanent structure with a life of 50 years. For timber used in falsework with a much shorter time under load these stresses are unduly conservative and may be considerably modified.

Table 1.3 gives the stresses which should be used in falsework design calculations provided the conditions listed apply. The minimum modulus of elasticity should be used where there is no load sharing.

Table 1.2 Equivalent strength classes for softwoods graded in accordance with BS 4978

Standard name	Strength class		
	SC3	SC4	SC5
Imported			
Douglas fir-larch (Canada)	GS	SS	—
Douglas fir-larch (USA)	GS	SS	—
Hem-fir (Canada)	GS, M50	SS	M75
Hem-fir (USA)	SS	—	—
Parana pine	GS	SS	—
Pitch pine (Caribbean)	—	GS	SS
Redwood	GS, M50	SS	M75
Southern pine (USA)	GS	SS	—
Spruce-pine-fir (Canada)	SS, M50	M75	—
Western red cedar	—	—	—
Western whitewoods (USA)	—	—	—
Whitewood	GS, M50	SS	M75
British grown			
Corsican pine	SS, M75	—	—
Douglas fir	GS, M50	SS	M75
European spruce	M75	—	—
Larch	GS	SS	—
Scots pine	GS, M50	SS	M75
Sitka spruce	M75	—	—

Note – Machine grades MGS and MSS may be substituted for the GS (general structural) and SS (special structural) grades respectively.

Table 1.3 Permissible stresses and moduli of elasticity for general falsework applications[1]

Strength class	Bending stress parallel to grain (N/mm²)	Tension stress parallel to grain (N/mm²)	Compression stress perpendicular to grain (N/mm²)	Shear stress perpendicular to grain (N/mm²)	Modulus of elasticity, E	
					mean (N/mm²)	minimum (N/mm²)
SC3	6.05	3.61	2.63	1.34	7344	4838
SC4	8.67	5.13	2.87	1.34	8035	5616
SC5	11.87	6.99	3.34	1.75	9418	6134

Load sharing[1]

Where a single member may carry its full share of the load unaided by its neighbours, the appropriate working stresses and the minimum value of modulus of elasticity should be used, but if members are spaced not more than 600 mm apart, and there is adequate provision for lateral distribution of loads by means of decking or joists spanning at least three supports, stresses may be increased by a factor of 1.1.

For the calculation of deflection of load-sharing systems, the mean value for the modulus of elasticity should be used.

Permissible stresses. The stresses given in table 1.3 may be used for calculations if the following conditions apply:

(a) the timber has been accepted as appropriate to the class concerned;
(b) sizes for dry timber are used;
(c) the duration of the load does not exceed one week;
(d) there is no wane;
(e) the bearing length does not exceed 75 mm, there is at least 75 mm of timber each side of the bearing, and take-up is not critical,
(f) the depth-to-breadth ratios of table 1.4 are not exceeded.

Table 1.4 Maximum depth-to-breadth ratios[1]

Degree of lateral support	Maximum depth-to-breadth ratio
No lateral support	2 : 1
Ends held in position	3 : 1
Ends held in position and member held in line, e.g. by purlins or tie-rods	4 : 1
Ends held in position and compression edge held in line, e.g. by direct connection of soffit formwork	5 : 1
Ends held in position and compression edge held in line, e.g. by direct connection of soffit formwork together with adequate bridging or blocking spaced at intervals not exceeding six times the depth	6 : 1
Ends held in position and both edges held firmly in line	7 : 1

The load/span graphs for timber sections in appendix A (fig. A.3) may be used in formwork design calculations where load sharing is applicable. Table A.1 gives the appropriate coefficient for three grades of timber and either a simply supported or continuous U.D. loading condition.

Timber which has been stress graded in the UK will be marked for identification. If it has been machine graded it will have the BSI kitemark. If it has been visually graded it will bear the TRADA mark to signify that the grader is certified by them. Gradings of the most commonly imported timber can usually be converted to their BS equivalent strength class.

There is an obvious problem with timber which has not been stress graded or has been used. Commercially graded timber will generally be acceptable as SC3 provided an inspection is made to reject sub-standard pieces. Stress-graded timber for re-use should be carefully inspected to ensure that it has not been ripsawn or has defects in excess of those permitted for the grade, otherwise it should be regraded.

When timber is used as a structural member in permanent work it is standard procedure to ensure that it is in accordance with the designer's specification. Unfortunately this is not always true with regard to temporary works particularly if site personnel are not familiar with the concept of strength classified timber. This problem can be overcome if (a) the temporary works designer ensures that the drawings for site use include a full specification for the materials to be used, including timber; (b) site management carry out a check on materials as an integral part of the supervision of temporary works construction. Site supervision procedures are discussed more fully in chapter 4.

Hardwood

The strength classifications in BS 5268 include hardwoods for structural use. Although not generally used for main members in falsework its hardness and ability to withstand impact makes hardwood a useful material for load-bearing wedges and packings, also for load-spreading caps which distribute point vertical loads to prevent punching into softwood bearers. It may also be shaped for special features more satisfactorily than softwood.

The most commonly stocked suitable hardwoods are keruing, Japanese oak, European beech, iroko and Burmese teak.

Plywood

Two basic grades of plywood are manufactured, interior and exterior. For formwork only exterior grade bonded with weather and boil proof adhesive (WBP), such as phenol formaldehyde resin, is suitable. As plywood is a factory product made under controlled condititions to recognised standards the allowable stresses and section properties quoted by the manufacturers trade organisations may be used with confidence by the formwork designer.

A variety of woods are used, depending on availability in the country of origin. Finnish birch faced plywood is made from a large number of plys, up to 19, with the interior plys alternatively birch and softwood. American and Canadian plywood is commonly faced with Douglas fir, with a variety of coniferous woods being used for the interior plys. British made plywood used for formwork is frequently faced with an African hardwood such as makore to give a durable and well-finished surface.

If plywood is not being marketed specifically for use in formwork advice should be sought regarding its suitability as some woods are known to create serious problems due to their interaction with concrete.

A description of the characteristics of the untreated face is used to specify quality, and hence cost, of the different grades of plywood. Douglas fir plywood varies in quality from (a) sanded both sides with defects cut out and neatly made good with wood or synthetic patches, to (b) unsanded sheathing with permissible knotholes and other minor defects left in the face.

Finnish birch faced exterior plywood is unsanded, the quality generally used in formwork has the larger knots replaced with tight plugs. Provided reasonable care has been taken with the edges the higher cost of having two faces prepared will be more than recovered by turning the sheet when one side is worn out.

Economic panel design is based on a module of the standard sheet sizes of 1200 × 2400 or 1220 × 2440 or 1525 × 3050. Larger sheets are available if necessary, the only real limitation to size is created by problems in transport and handling.

Although 12 mm and 18 mm are the most common thicknesses used a range of other thicknesses, from 6.5 mm used for lining and forming curves to 27 mm used for heavy construction, is available.

Plywood may be used plain with the surface untreated except for the application of a releasing agent for ease of striking. The surface produced is generally acceptable for unexposed surfaces although fir plywood produces a pronounced grain pattern after several uses. For improved surface quality and re-use potential a variety of surface coatings are available (table 1.5). Barrier paints and varnishes can only be applied to clean dry surfaces with a moisture content below 15%. Ideally this should be done under factory conditions by the manufacturer. Two or three coats are required, any cut edges should be sealed with a water-proofing agent such as aluminium or chlorinated rubber paint.

Table 1.5 Concrete finish – birch-faced plywood

Plywood	Concrete finish
Regular	High quality surface not exposed to view but free from surface irregularities or Medium quality exposed surfaces. If overall uniformity of colour required pre-treat with wax or polyurethane paint
Phenol, epoxy or polyurethane coated or painted plywood (applied during manufacture)	High quality exposed surfaces with overall uniformity of colour
Film-faced plywood	High quality exposed surfaces with overall uniformity of colour
Phenol coated or film faced with textured surface	Textured high quality concrete surface to be exposed to view. Overall uniformity of colour
Profiled plywood with polyurethane coat	Profiled high quality exposed concrete surfaces. Overall uniformity of colour
GRP surfaced plywood – smooth coat, textured or patterned surface	High quality exposed concrete surfaces with overall uniformity of colour Smooth, textured or patterned surfaces obtainable

It has been belatedly accepted by all sides of the industry that it is next to impossible to produce large areas of plain concrete surface to a satisfactory aesthetically pleasing standard. Patterned, profiled and textured coatings on plywood are available and produce a variety of interesting surfaces which effectively tolerate surface blemishes and minor colour irregularities and can make an eye catching decorative feature.

Plywood is a strong constructional material but it is expensive. Up to 60% of the total cost of a concrete structure may be in formwork so the incentive exists for the formwork designer to employ plywood economically by fully utilising strength properties, surface quality and re-use potential (table 1.6).

Table 1.6 Plywood re-use potential[15]

	Re-uses
Untreated (except for release agent)	Up to 10
Treated with barrier paint/varnish	Up to 20
Factory processed resin impregnation	Up to 50

Timber generally, and plywood in particular, tends to be regarded by site personnel as a consumable item rather than an item of plant which will repay the cost of maintenance and careful use by an extended economic life. Temporary works material, including timber and plywood, should be regularly inspected, suitably identified and marked, and carefully stored.

Glass fibre reinforced plastic (fibre glass)

Panels and moulds for precast and *in situ* concrete are made by building up a sandwich of glass fibre mat and epoxy or polyester resin onto a pattern of the finished shape. The thickness will vary from 5 mm to 16 mm depending on the strength and stiffness required, timber or metal inserts can be incorporated for additional reinforcement. The manufacture is generally left to specialist firms as the process requires controlled conditions.

Although grp is a strong lightweight material it is too flexible for use in large flat areas and is most suitable for repeated complex shapes or indented surfaces where a high standard of finish is required. It is commonly used for decorative feature panels and structural elements like waffle and trough floors or flared columns for which the only practical alternative would be moulded polypropylene or more expensive sheet steel.

Unreinforced plastics are generally too flexible for formwork use.

Aluminium

Extruded aluminium sections are now widely used in proprietary formwork. Lightweight beams and soldiers are made in this material, also panel framing.

1.2 SURFACE FINISHES

In addition to the surface finishes produced by different types of plywood there are a variety of techniques and formwork linings used to create special effects and decorative concrete features both internally and externally.

Timber

Sawn timber boards may be used to produce an interesting textured surface. Straight markings will be made by using a band saw and radial markings by a circular saw. Bold relief can be obtained by using boards of varying thickness. Boards of random width are also used.

Even with a textured surface it is advisable to break up large areas into panels or bands, ensuring that the position of tie rods does not detract from the required effect.

The dimensions should be chosen with regard to the scale of the work and the overall appearance of the structure.

Board marked concrete surfaces have been used to create very successful architectural features especially when emphasized by effective lighting (Plates 1–3). The cost and difficulty should not be under estimated, very careful design and site supervision are required.

Exposed aggregate

The attractive colour and graining of natural stone is displayed as a decorative feature in exposed aggregate finishes. Three methods are used.

(a) Aggregate transfer. Aggregate in the form of single size stones or pieces of crushed rock, usually 50 mm or larger, is hand placed and glued to the shuttering face. The surface of the embedded aggregate is exposed when the shutter is stripped and the glue matrix removed (Plate 4).

(b) Bush hammering. The top surface of concrete is chipped away using a percussion tool to expose a cross-section of concrete. No special aggregate is used, the effect is created by the natural colour and texture of the concrete materials.

(c) Brushing. In this method aggregate is exposed by spraying and brushing to remove fines and reveal the stones. Selected stones may be placed against the face shuttering if the work is to be cast face down. A retarding agent applied to the shutter will assist in the removal of cement grout (Plates 5 and 6).

Although all these methods may be applied to *in situ* concrete high quality finish is best achieved by using precast panels manufactured under controlled conditions.

Glass fibre reinforced plastics

Plate 7 shows the use of a deeply textured surface obtained by using grp form liners in the treatment of a large exterior concrete area. Carved *polystyrene* liners were used for the intermittent sculptural panels. GRP relief panels were also used to produce the feature interior walls shown in Plate 8.

Form liners

Moulds made in elastomeric sheets from an original surface such as rock, brick, stone, timber or textured concrete are used as form liners to create a variety of interesting concrete surfaces.

Thermoplastic is also used in a similar manner (Plate 9). Plates 10 and 11 illustrate the effective use of surface marking on large scale concrete work.

1.3   CUSTOM MADE AND SPECIAL PURPOSE FORMWORK

Slip form construction

Slip forming may be described as a vertical extrusion process in which the formwork serves as a moving die. Vertical rate of movement is controlled to ensure that the concrete is self-supporting when it emerges from the formwork. Specially designed inner and outer wall forms together with access and working platforms are supported by a number of hydraulic jacks which grip tubes or rods embedded in the middle of the wall. The whole assembly climbs upward on the jack rods in a continuous operation by progressive small bites of the jacks.

This method of construction is applicable to tall structures of uniform cross-section with no projections. Apertures in the finished work, doors, windows, flues and the like are formed by boxing out. Groups of tanks or silos with multiple use of the equipment are a common application. Chimneys, service cores in multi-storey buildings, and shaft linings are other suitable structures. Plate 12 shows a typical application of the method.

Permanent formwork
Plates 13 – 16 illustrate a variety of uses for permanent formwork. Expanded metal is particularly useful in confined locations with projecting reinforcement where formwork removal and surface preparation would be difficult.

Tunnels and culverts
The tunnel lining shutter in Plate 17 is fully telescopic with hydraulic operation. Shutter rings to completed concrete lining are collapsed, moved forward and re-erected. The concreting process is continuous with no joints in the lining. Plate 18 shows the double skin shutter with traveller for cut and cover construction of a motorway.

Precast units
Plate 19 shows the construction of units for an elevated roadway. The mould for bridge beams of standard cross-section is shown in Plate 20.

Purpose-made formwork for *in situ* concrete
Plates 21 – 23 illustrate a range of uses for special formwork. BS 5975 recommends that formwork should be designed to resist wave forces, Plate 21 shows a location where this will be necessary.

An interesting challenge presented to the formwork designer is the casting in concrete of complex water passages associated with large scale pumps, turbines and other hydraulic plant. Plates 24 and 25 illustrate different solutions to this problem. In the first method turbine draft tubes are formed by casting in steel liners fabricated to shape. The alternative method uses three-dimensional timber formwork as a former for the complex shape of turbine condenser water passages. Good practical detailing will be required to ensure that the timber can be dismantled and withdrawn.

Plates 26 and 27 indicate the types of complex construction which are suitable for purpose made formwork.

Plate 1 Patterned effect using boards of different grains

Plate 2 High quality finish using shot-blasted Douglas fir boards and metal tie rod inserts

Plate 3 Interior use of board marked *in situ* concrete with relief emphasised by lighting

Plate 4 Aggregate transfer finish on bridge pier using 35 mm crushed granite aggregate

Plate 5 Hand placed aggregate exposed by washing

Plate 6 8 mm graded aggregate exposed by water spray

Plate 7 Sculptured retaining wall using glass fibre and carved polystyrene form liners

Plate 8 Feature interior concrete finish produced by casting against repeated 6.4 mm relief grp forms

Plate 9 Detail of white concrete cast against textured thermoplastic

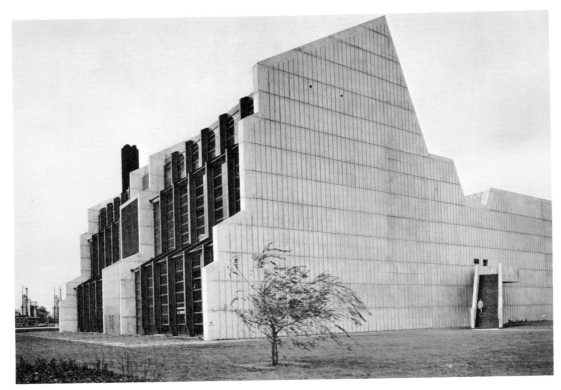

Plate 10 Concrete wall with large scale repeating relief, effective from some distance

Plate 11 Board marked panels with horizontal joint lines emphasised create an interesting architectural effect on an otherwise featureless building

Plate 12 Tall circular tower constructed by slip form method

Plate 13 Large scale use of expanded metal as permanent formwork to construction joints

Plate 14 Difficult boxing-out for anchorage units formed in expanded metal

Plate 15 Aircraft shelters with permanent steel liners and curved external shutters

Plate 16 Large decorative precast concrete panels being used as permanent formwork

Plate 17 360° telescopic tunnel lining shutter

Plate 18 Double skin arch formwork

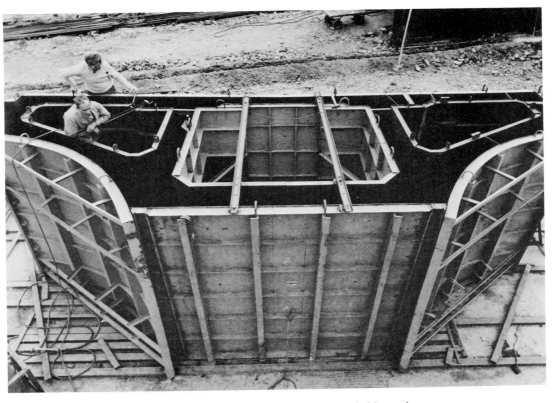

Plate 19 Mould for post-tensioned precast bridge unit

Plate 20 Steel mould for standard prestressed bridge beam

Plate 21 Formwork for fender pile cap

Plate 22   Crosshead shutter to elevated roadway

Plate 23   Formwork for circular tank with integral overflow weir and collecting channel

Plate 24 Hydroelectric turbine draft tube sections

Plate 25 Steam turbine condenser outlet passage and manifold

Plate 26 Internal and external arch formwork for fertiliser plant

Plate 27 Trial erection of single carriageway falsework for viaduct

Chapter 2
Column and Wall Formwork

2.1 COLUMNS

Columns tend to occur in either considerable numbers with identical dimensions or few in number but having special architectural features. In either case considerable attention will be directed to the arrangement and design of the formwork. Multiple column construction with repeated use of the formwork will reward careful design with rapid progress and economy of operation. Columns which have been made an architectural feature, such as in the entrance hall of a public building, with a flared or sculptured shape, or special surface finish will usually require purpose made formwork.

The volume of concrete in a column is small compared to the rate at which concrete is supplied so columns are filled in a very short time and full hydraulic concrete pressure is developed unless the dimensions are small enough to create arching. Reinforced concrete columns which are part of the structural frame of a building are generally constructed to the full storey height in one operation, up to the underside of intersecting beams or the soffit of the next floor. If it is practical to bend up projecting lap reinforcement for intermediate beams or floors storey height need not be a limitation. Formwork for columns of constant cross-section can be made continuous over two or more floors.

Bridge piers and the like, unless exceptionally tall, are constructed in one piece, the only limit to height being dictated by either economy or problems in placing and vibrating concrete.

2.1.1 Column formwork components

Kicker

Kicker formwork for straight sided columns is usually made from dressed timber 50 mm high made into a frame which is accurately positioned to the setting out lines and shot bolted or otherwise fixed to the foundation or floor slab. Circular and other curved shapes are formed in cut out plywood ribs and timber spacers with the aperture corresponding to the column section lined in hardboard or thin plywood.

Plywood and timber shutters

19 mm plywood is generally used for column sheeting. The plywood sheets are either cut to suit column dimensions and stiffened with vertical timber studs, or framed with timber into standard panels for convenience in re-use and greater versatility. For practical reasons unframed plywood is generally used so that it spans with the grain of the external plies in the weakest orientation. This should be allowed for in design calculations.

Metal forms

Proprietary steel framed forms with either sheet steel or plywood facing are manufactured in a variety of sizes suitable for use as shutters for columns of standard
dimensions. Panel sides are connected using wedge clamps to external corner sections
incorporating great rigidity to the box section, particularly if panel joints and external
corner joints are staggered (fig. 2.1).

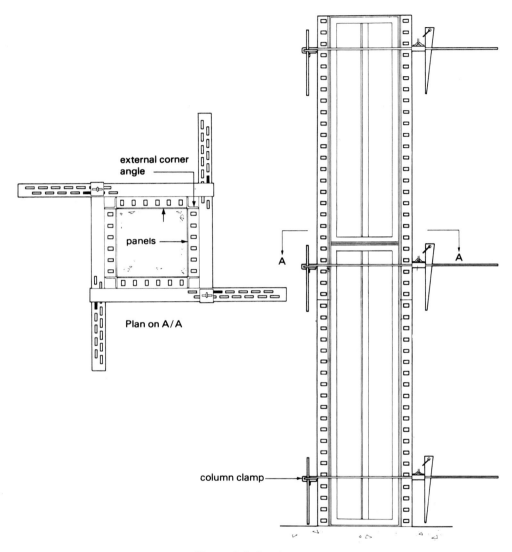

external corner
angle

panels

Plan on A/A

A   A

column clamp

Figure 2.1 Steel panels

Column clamps

Proprietary column clamps, available from the major construction accessories suppliers,
are now widely used. They are made from 57 mm by 8 mm mild steel in a range of
compatible sizes. Slotted holes and wedge fixings allow for adjustable lengths up to a
maximum of 1220 mm.

Column clamps are extremely useful and safe items of equipment provided they are
not used indiscriminately. They are intended for light construction with a spacing of
about 225 mm at the recommended maximum concrete pressure of 50 kN/m². In
practice if plywood and timber are of reasonable size the clamp spacing may be

calculated with regard to shuttering deflections only, normally no clamp calculation is necessary. Adjustable column clamps are designed to resist the tensile and shear forces which arise when the clamps are used to tie the corners of the column shutter, they are not designed to act as walings. However, it is common and acceptable practice to introduce an intermediate vertical stud to reduce the span of the plywood facing (fig. 2.2).

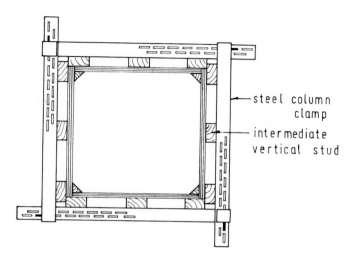

Figure 2.2 Steel column clamp

Column box

Although steel clamps are now universally used for small columns an alternative is to make up a column box in plywood facing with a yoke in timber and mild steel external tie bolts (fig. 2.3). For ease in erection and re-use the box sides are fabricated into large panels. Folding wedges may be introduced for rapid striking. A variety of irregular column sections can be superimposed on the basic box shutter.

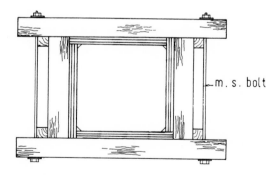

Figure 2.3 Timber yoke and bolts

Large columns

Where tie rods are not permitted for aesthetic reasons yokes fabricated from steel sections are required to support timber and plywood panels. If through ties are used, formwork for column sides can be designed as walls using the methods described in the following section. Adjacent walings should be positively connected to tie in the column corners and prevent leakage of grout; a simple slot and wedge arrangement is suitable.

Most proprietary heavy duty soldiers may be used as walings for large columns. Cantilevered shutters are used in the construction of very tall columns where the cost of full height formwork could not be justified or would not be practical.

External bracing
Formwork for tall slender columns has little resistance to twisting, external props are required to align the formwork initially and keep it vertical and true during concreting. Care must be taken as any twisting or deviation from vertical will be obvious in the finished work. The use of inclined props for bracing will result in an upward component of force when the prop is loaded. To counteract any tendency towards upward movement the bottom of the formwork should be wedged tightly to the kicker, or held down by some other means. Adjustable push–pull steel props attached at a corner and fixed to the floor slab are particularly effective.

2.2 WALL FORMWORK

Developments in formwork and falsework for building and civil engineering reflect the trends in construction methods. Spiralling costs of overheads and labour have accelerated the introduction of falsework systems which can be rapidly erected with the minimum use of on-site craftsmen. Very large lifts are now commonplace with the number of time-consuming wall ties being drastically reduced by the use of heavy duty soldiers. The use of aluminium has reduced the weight of formwork panels and soldiers. Progress in this area has been matched by developments in construction equipment, notably high capacity concreting and lifting plant.

In the industry extremes in formwork style and scale co-exist quite happily; from complex custom-made schemes designed and fabricated by the supplier to simple rule-of-thumb site-made shuttering. In between lies a variety of formwork which reflects the huge range of construction activities.

The formwork designer's task is to design the most economical scheme with the resources available and within the overall construction programme. The general structural principles and arrangement of components will be the same regardless of the materials used. However, it should be noted that formwork calculations using standard timber and mild steel sections are based on accepted working stresses. If manufactured products are incorporated, or used as a complete system, full information regarding technical capabilities and method of use should be obtained from suppliers before deciding on permissible working loads.

2.2.1 Wall formwork components

Kicker
Except for the lowest grade of work wall construction is commenced by the formation of a starter or kicker, 50 mm to 75 mm high, which is set accurately and used as a guide and positive fixing point for subsequent wall formwork. The kicker is usually cast in a separate operation on top of the foundation or floor slab, alternatively it may be possible to suspend a kicker box or frame at the correct line and level and cast it integrally with the foundation. This latter method is particularly useful in water-retaining structures where it is convenient to set the horizontal water bar so that it projects from the kicker upstand. To prevent grout loss the first lift of wall formwork is held securely on to the kicker by external wedging or internally by tie rods set as low as possible. This will also resist any tendency for the formwork to lift either through the use of raking shores or normal vibration. However, if these forces are thought to

be significant it is much better practice to tie the formwork down by using anchors buried in the kicker or foundation concrete.

Plywood panels

Despite the increasing use of proprietary steel and aluminium framed panels in wall construction the ubiquitous plywood faced timber panel is still the universal shuttering component. Panels range in sophistication from standard plywood sheets framed with nailed sawn timber, to complete systems using factory made panels and accessories in a range of compatible sizes. The plywood face screwed to a jointed dressed timber frame will usually be reversible to increase the economic life of the panel. The frame is painted and the face treated with polyurethane or some other finish. Purpose made panels can be as large as required but stock panels are usually small enough to be man handled. If necessary wall shuttering can be broken down into its component parts between lifts. Usually lifting gear is available, panels are bolted together into larger sections and lifted complete with soldiers and walings attached. Corners are a potential weak point in any wall formwork arrangement. Most contractors' panel systems will include special external and internal corner units which are bolted to adjacent wall panels and effectively eliminate this problem.

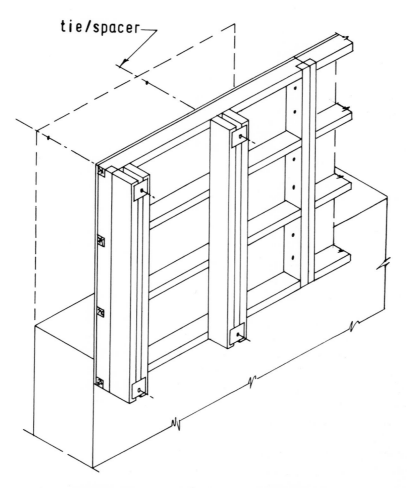

Figure 2.4 Plywood and timber panel typical arrangement

Wall ties and anchors

Simple re-usable wall ties are made from mild steel rod threaded at both ends and fitted with nuts and plate washers. The washer/bearing plate should be big enough to reduce the bearing stress on timber to less than 2.75 N/mm². It is possible to withdraw greased or oiled rods if they are turned while the concrete is green to break the bond. A more satisfactory method is to insert the rod into a disposable cardboard or plastic tube.

Anchors for use in single-sided wall construction can be made by embedding a standard nut welded to a plate washer in the concrete near the top of a lift. The size of plate and position of the anchor are determined by the pull out strength required, this in turn is determined by the shear strength the concrete has developed by the time the anchor is loaded. The anchor nut is located in position during concreting by means of a tapered dumb bolt fixed through the wall formwork. An alternative type of anchor is made by welding a loop or bent bar to the nut (fig. 2.5).

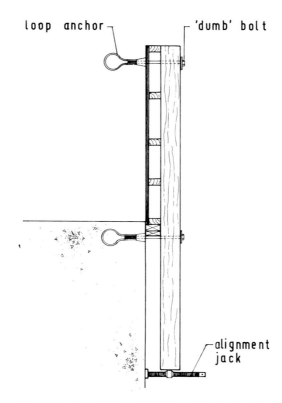

Figure 2.5 Cantilevered formwork for single-sided walls

It is recommended that a load factor of at least 2.5 against failure is used for ties and a factor of safety of 3 for anchors when resistance of concrete to pull out is considered.

A variety of proprietary wall ties and anchors are available. Some are designed for use as part of a specific wall formwork system but many are suitable for general use (fig. 2.6).

Walings and soldiers

Conventional walings and soldiers are made up using either stock timber or mild steel beam and channel sections. The most common arrangements are illustrated in figs. 2.4, 2.5 and 2.10.

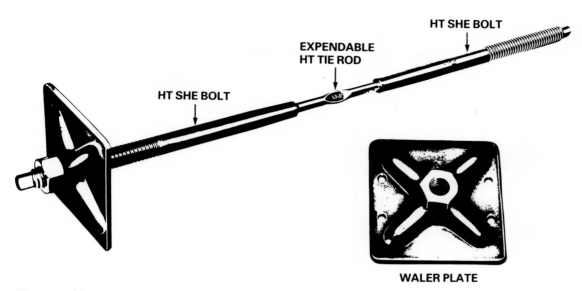

HT SHE BOLT

EXPENDABLE HT TIE ROD

HT SHE BOLT

WALER PLATE

Figure 2.6(i) High tensile tie for medium/heavy duty work. The crimped area at the centre of the expandable tie rod prevents rotation when the She bolts are being removed from finished concrete. This type of tie is particularly suitable for large crane handled panels as the assembly can be passed through from one side

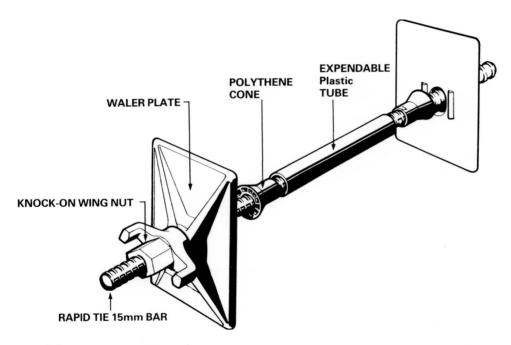

POLYTHENE CONE

EXPENDABLE Plastic TUBE

WALER PLATE

KNOCK-ON WING NUT

RAPID TIE 15mm BAR

Figure 2.6(ii) High working load tie made from deformed high tensile steel reinforcement bar. All parts are recoverable except for the PVC tube which allows the tie bars to be extracted

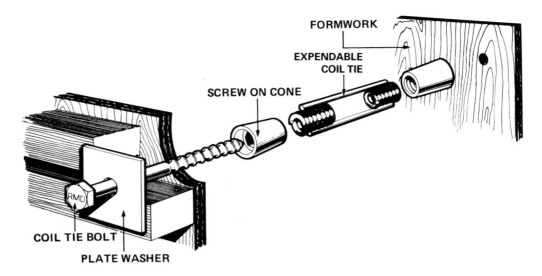

Figure 2.6(iii) Coil tie system. Round nuts at each end of the expendable centre section are made from coiled wire. The screw on plastic cones also serve as distance pieces so the complete assembly acts as a wall spacer to maintain correct thickness

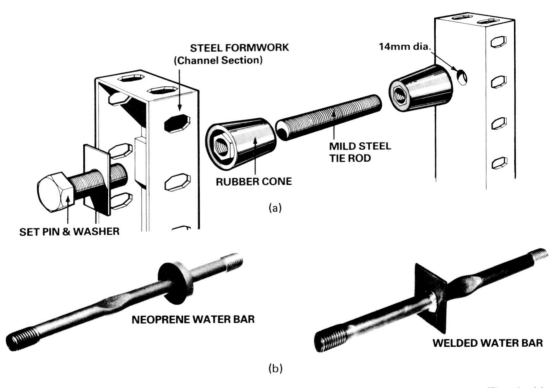

Figure 2.6(iva) The action of this mild steel tie system is similar to the previous. The double nut connecting parts of the threaded tie bar is moulded into a rubber cone for ease of removal. **(ivb)** For use in water-retaining structures the tie bar can be fitted with a water bar

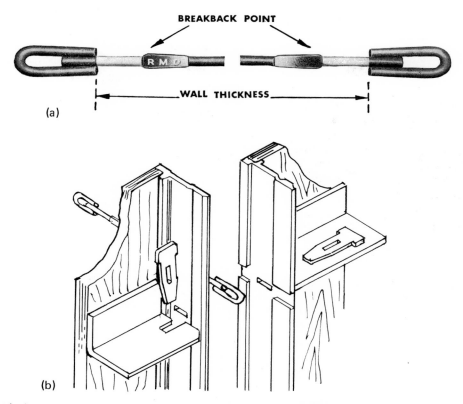

Figure 2.6(va) This snap tie is part of a system using steel framed panels. The tie is weakened 25 mm from the face by crimping. After concreting the outer part of the bar is broken off by twisting and the hole plugged. **(vb)** Shows the wedge assembly which simultaneously joins the panels and fixes the snap tie in position

LOOP COIL TIE   CONE   PLATE   COIL TIE BOLT
                                 WASHER

Figure 2.6(vi) Loop coil anchor tie used with single faced or climbing formwork

Proprietary walings and soldiers are increasingly versatile, many can be used with a variety of facing panels, and also as components in floor centering. Their use is shown in fig. 2.7.

Standard mild steel scaffold tubing is employed by some manufacturers for walings. It is easily bent to radius and is particularly useful for curved walls (fig. 2.9(ii)). Although scaffold tubing can be used generally its best performance is achieved when used with the steel framed panels and ancillary items specially designed for it.

2.3   PROPRIETARY WALL FORMWORK COMPONENTS

2.3.1   Proprietary soldiers

A wide range of versatile soldiers is available from formwork suppliers. They are made from pressed steel or extruded aluminium sections fabricated into back-to-back twin

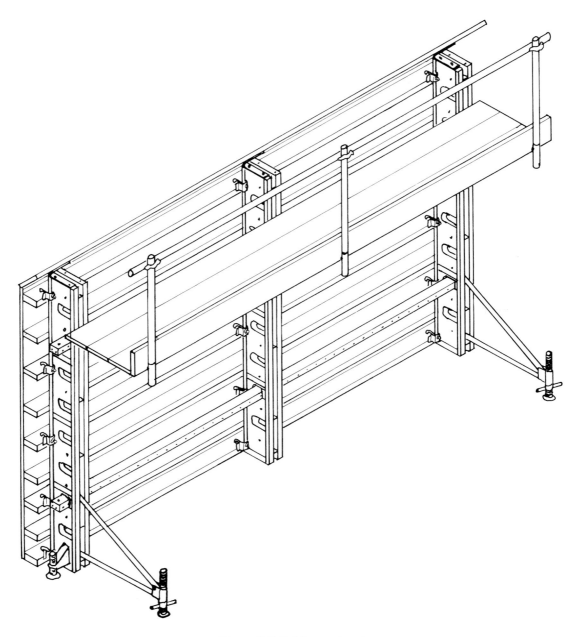

Figure 2.7 Soldier system

units for ease in positioning tie bolts. Primarily they are intended to replace twin 150 × 75 and 225 × 75 timber soldiers in medium duty work and to replace rolled steel channels and joists in heavy duty work. Like the timber sections they replace they can also be used as horizontal walings to wall and column formwork and as primary and secondary beams in slab centering. A useful selection of ancillary fittings complement the soldier units; lifting points, access brackets, plumbing feet, waling clamps, all are designed to make it easy to assemble and use formwork in large panels with a variety of waling and panel materials. Very large forms are made by joining soldier units with butt plates and splices. The reduced moment or resistance at the joint can be accommodated by careful placing of tie bolts.

Table 2.1 gives brief details of a selection of proprietary soldiers. Full information for design should be obtained from the manufacturers technical specification.

Table 2.1 Proprietary soldiers – characteristics

(i) Acrow open web

lengths: 900, 1800 and 3000
jointing: channel splice
weight: 17.5 kg/m
max. working bending moment: 23 kNm
max. tie load: 105 kN
formwork: panel system and timber

(ii) Mabey Mk II

lengths: six No. 675–4050
jointing: bolted end plates
weight: 21 kg/m
max. working moment of resistance: 25.3 kNm
max. working tie load: 100 kN
formwork: timber

(iii) Kwikform aluminium

lengths: four No. 2700–6000
jointing: bolted jointing unit
weight: 10.47 kg/m
max. tie load: 160 kN
formwork: aluminium walings

Table 2.1 (*contd*)

(iv) RMD super slim 2

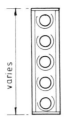

lengths: 900, 2700 and 3600
jointing: bolted end plates
weight: 19.75 kg/m
working bending moment: 31–38 kNm
working tie load: 90–120 kN
formwork: panel system and timber

(v) SGB heavy duty strongback

lengths: 1500 and 2850
jointing: turnbuckle and splice plates
weight: 23 kg/m
moment of resistance: 55.9 kNm
max. tie load: 118 kN
formwork: system steel walings and timber

(vi) GKN aluma strongback

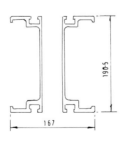

lengths: 2750, 3809 and 4876
jointing: splice plates
weight: 11.9 kg/m
max. permissible bending moment: 34 kNm
max. permissible tie load: 175 kN
formwork: aluminium walings and timber

2.3.2   Proprietary wall formwork systems

The basic unit of wall formwork systems is the panel. These may be fabricated with high tensile steel frames and strengthening ribs and either sheet steel or inset plywood facing. Twelve millimetre plastic faced plywood is now the most commonly used material. Panels are also made from aluminium alloy extrusions with a facing in 15 mm film faced plywood. Aluminium panels are about 40% lighter than the equivalent steel panels.

Panels are usually joined with a simple knock-out wedge and key which may incorporate a fixing for snap ties (see fig. 2.6(v)).

For maximum versatility systems panels are made in a variety of sizes for use in columns, beams, slabs and walls. Wall systems are basically designed for use with snap ties and scaffold tube walings (fig. 2.8). However, for heavy duty work with large crane handled panels most systems can be adapted to use high capacity conventional wall ties and the manufacturer's proprietary soldiers.

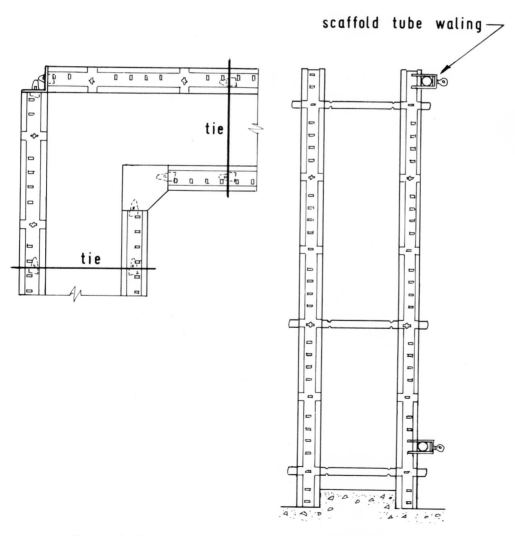

Figure 2.8 Aluminium, plywood and snap tie wall formwork system

Aluminium/plywood
Figure 2.8 shows a proprietary system with aluminium and plywood panels, snap ties and scaffold tube walings for alignment.

Steel panels
Figure 2.9 shows a steel panel system in a variety of uses. Heavy duty soldiers are incorporated (iii), for cantilever use in single wall construction.

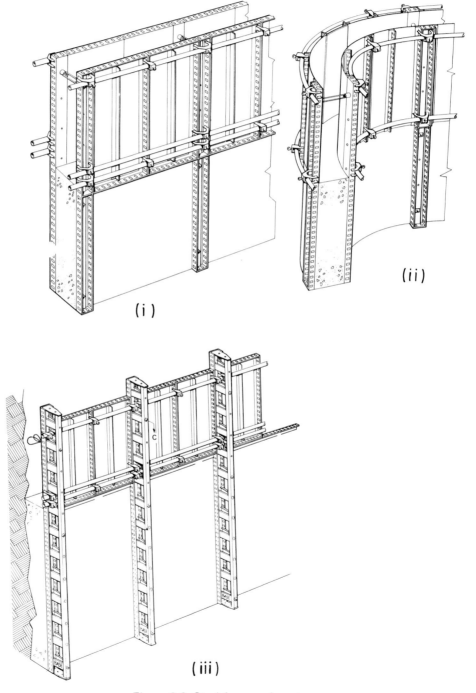

(i)

(ii)

(iii)

Figure 2.9 Steel formwork system

A wide variety of solutions are possible for column and wall formwork. A selection of methods and materials is used in the formwork arrangements illustrated in Plates 28–36. Plate 37 shows typical *in situ* multi-storey construction.

2.4   CONCRETE PRESSURES

The effect of high frequency vibration on freshly placed concrete within formwork is to keep it in a fluid state so that it behaves almost as a liquid of the same density subject to the normal laws governing hydrostatic pressures. Full hydrostatic pressure may or may not be developed depending on whether stiffening or arching of the concrete occurs before the lift is finished. Which of these limiting conditions will apply, height limit, arching limit or stiffening limit, depends on a number of factors.

1. Density of the concrete, Δ (kg/m³)
2. Workability of the mix, slump (mm)
    Workability may be classified in terms of the standard slump test. It is dependent on the mix proportions, water/cement ratio and the shape and grading of the aggregates. Concrete of low workability will stiffen much more rapidly than 'wet' concrete with a large slump.
3. Rate of placing, R (m/h)
    When concrete starts to stiffen hydrostatic pressure on the formwork immediately starts to die away, only fresh concrete above the level of stiffening will exert pressure. If the concrete sets before the lift is completed full hydrostatic pressure will not be developed, a reduced pressure, the stiffening limit, will apply.
    Will the formwork be filled before the pressure is limited by the setting effect of high temperature and low workability? This is dependent on the vertical rate of placing which is calculated from the rate of concrete delivery and the plan area of the lift.
4. Concrete temperature, (°C)
    Since the rate of the chemical reaction is governed by temperature the time taken to stiffen will decrease as concrete temperature increases.
5. Height of lift, H (m)
    Maximum hydrostatic pressure (P_1) will be developed when the full depth of concrete in the lift remains fluid. It may be calculated from the expression:

$$P_1 = \frac{\Delta H g}{1000} \text{ kN/m}^2$$

    where g is the gravitational constant to convert mass to force (9.81 m/s). Alternatively, for concrete density of 2500 kg/m³ hydrostatic pressure may be taken to be 25 kN/m² per metre of height.
6. Minimum dimensions of the section cast, d (mm)
    In narrow sections during compaction the aggregate develops an arching or bridging effect which is capable of supporting a surcharge of fresh concrete and limits the development of full hydrostatic pressure. This reduced pressure, arching limit, may occur if the least dimension of the section is 500 mm or less.

The combined effect of these variables on ultimate concrete pressure was investigated by the Construction Industry Research and Information Association. The original work, published as CIRIA Report 1, has been simplified into a useful data sheet, part of which is used here by kind permission of the Director (table 2.2).

Table 2.2 Concrete pressure on formwork. (Based on the CIRIA Data Sheet 'Concrete pressure on formwork', by permission of the Director of the Construction Industry Research and Information Association.)

The factors to be considered are

This data sheet is based on CIRIA Report 1 and is applicable only to Portland cement concretes without cement replacement materials or admixtures, and to compaction by internal vibration.

1. Density of the concrete (kg/m^3)
2. Workability of the mix, slump (mm)
3. Rate of placing, R (m/h)
4. Concrete temperature (deg. Celsius)
5. Height of lift, H (m)
6. Minimum dimensions of the section cast, d (mm)

Design pressure P_{max}

The design pressure is limited by the height of lift, arching of the concrete, or stiffening of the concrete with an overall practical limit taken to be 150 kN/m^2. The limiting values (P_1, P_2, P_3) are given below (to the nearest 5 kN/m^2) and the design pressure is taken as the least of these. An allowance for impact arising from placing operations is not normally necessary. Examples are given overleaf.

1. Height limit (hydrostatic pressure taken to be 25 kN/m^2 per metre of height).

H (m)	1	2	3	4	5	$\geqslant 6$
P_1 (kN/m^2)	25	50	75	100	125	150

2. Arching limit

d (mm)	R (m/h)											
	1	2	3	4	5	6	8	10	15	20	30	$\geqslant 40$
150	P_2 =35	35	40	45	45	50	55	60	75	90	120	150
200	40	40	45	50	50	55	60	65	80	95	125	150
300	50	50	55	60	60	65	70	75	90	105	135	150
400	60	60	65	70	70	75	80	85	100	115	145	150
500	70	70	75	80	80	85	90	95	110	125	150	150

3. Stiffening limit

Slump (mm)	Concrete temp.(°C)	R (m/h)									
		1	1.5	2	2.5	3	4	5	6	7	$\geqslant 8$
50	5	P_3 =50	70	95	115	135	150	150	150	150	150
	10	40	55	70	85	100	135	150	150	150	150
	15	40	45	55	65	75	100	125	150	150	150
	20	35	40	45	50	55	70	90	105	125	150
75	5	60	85	110	140	150	150	150	150	150	150
	10	50	65	85	105	125	150	150	150	150	150
	15	40	50	65	80	95	125	150	150	150	150
	20	35	40	50	60	70	90	115	135	150	150
100 to 150	5	70	100	130	150	150	150	150	150	150	150
	10	55	75	100	120	150	150	150	150	150	150
	15	45	60	75	90	110	150	150	150	150	150
	20	35	45	55	70	80	110	130	150	150	150

Examples

A. Column: 300 × 300 mm × 3 m high

Rate of placing: 15 m/h

Slump: 100 mm

Concrete temperature: 10°C

Height limit, P_1 =   75 kN/m²

Arching limit, P_2 =   90 kN/m²

Stiffening limit, P_3 =   150 kN/m²

The design pressure is the lowest of the above three values

∴Design pressure, P_{max} = 75 kN/m²

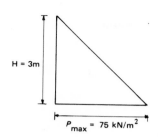

H = 3m

P_{max} = 75 kN/m²

B. Wall: 150 mm thick × 4 m high

Rate of placing: 3 m/h

Slump: 75 mm

Concrete temperature: 15°C

Height limit, P_1 =   100 kN/m²

Arching limit, P_2 =   40 kN/m²

Stiffening limit, P_3 =   95 kN/m²

∴Design pressure, P_{max} = 40 kN/m²

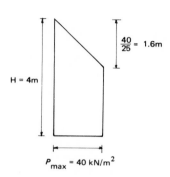

H = 4m

$\frac{40}{25}$ = 1.6m

P_{max} = 40 kN/m²

C. Wall: 500 mm thick × 6 m high

Rate of placing: 2 m/h

Slump: 50 mm

Concrete temperature: 15°C

Height limit, P_1 =   150 kN/m²

Arching limit, P_2 =   70 kN/m²

Stiffening limit, P_3 =   55 kN/m²

∴Design pressure, P_{max} = 55 kN/m²

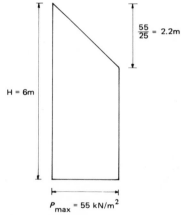

H = 6m

$\frac{55}{25}$ = 2.2m

P_{max} = 55 kN/m²

D. Mass concrete wall: 2.4 m lift height with inclined cantilevered form

Rate of placing: 600mm/h

Slump: 50 mm

Concrete temperature: 5°C

Height limit, P_1 =   60 kN/m²

Arching limit, P_2 Inapplicable, d > 500 mm

Stiffening limit, P_3 =   50 kN/m²

∴Design pressure, P_{max} = 50 kN/m²

Note: Pressure acts at right angles to form face.

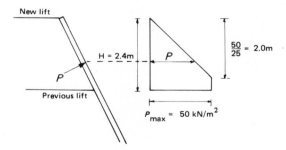

New lift

Previous lift

H = 2.4m

$\frac{50}{25}$ = 2.0m

P_{max} = 50 kN/m²

2.5   WALL FORMWORK CALCULATIONS – EXAMPLES

2.5.1   Example 1

Design of formwork for a reinforced concrete abutment wall. Abutment wall is 800 mm thick, 7.5 m high, in bays 9.5 m long.

Data: concrete supplied at the rate of 20 m³/h. 75 mm slump, normal vibration, maximum temperature 15°C.

Design brief: Use plywood, sawn timber and m.s. tie rods. No heavy lifting gear available. A suitable arrangement is illustrated in figs. 2.10, 2.11 and 2.12.

Concrete pressure – Use first lift for design.
Vertical rate of placing, $R = 20/(0.8 \times 9.5) = 2.63$ m/h
  (i) $d > 500$ mm, arching limit not applicable
 (ii)   from table 2.2 stiffening limit = 83.9 kN/m²
(iii)   height limit 25×1.67 = 41.75 kN/m²

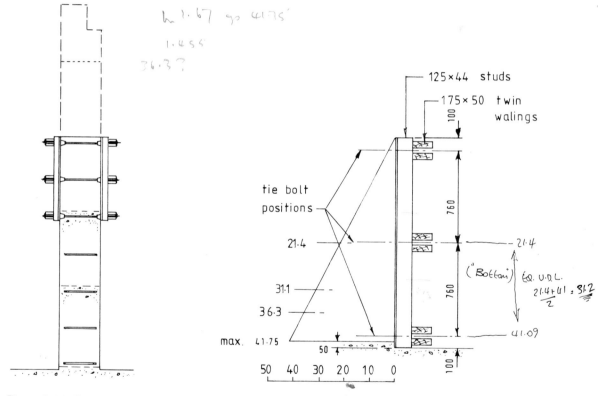

Figure 2.10 Formwork arrangement with lifts

Figure 2.11 Pressure diagram (kN/m²)

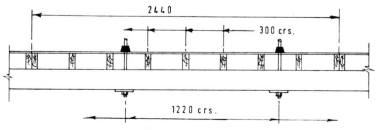

Figure 2.12 Plan

Plywood sheathing

p. 134.

Design pressure = 41.75 kN/m²

Using unsanded COFIFORM Douglas fir, table A.3 gives 20.5 mm, 7 ply, at 300 mm support spacing.

Vertical studs (bottom)

From pressure diagram equivalent UDL = 31.1 kN/m²   *SEE NOTE ON DIAG. 2.11.*

Total load = 0.76 × 0.3 × 31.1   = 7.09 kN

From safe load graph 2 (fig. A.3), 125 × 44 class SC3 timber is satisfactory.

Walings

Tie rods at 1220 mm centres.

Total load on bottom waling 36.3 × 1.22 × 0.43 = 19.04 kN

Total load on middle waling 21.4 × 1.22 × 0.76 = 19.84 kN

Load on single waling   = 9.92 kN

From safe load graph 3 (fig. A.3), 175 × 50 class SC3 timber is satisfactory, used in pairs.

Bolts

Maximum load on bolts is 19.84 kN, allow 20% additional for effect of waling continuity.

∴ Design load = 19.84 × 1.2 = 23.81 kN

Minimum diameter for 130 N/mm² stress = 17.5 mm, use 20 mm diameter to BS 4190, or acceptable alternative.

Washers

Plate washers are required to transfer tie rod loads on to the timber walings. Normally 6 mm mild steel plate is used, with a bearing area calculated to ensure that the compression stress in table 1.3 is not exceeded.

Panels

The calculation has assumed a panel size of 2440 × 1720, based on 1½ standard sheets. This will conveniently accommodate the bay length and allow a slight variation in height to follow deck camber. Expendable tie rods with rubber cones will be used.

The next step would be to prepare working drawings of panels and an assembly drawing.

Economics

The selection of lift heights and panel dimensions has been dictated by the method of erection; single panels and walings lifted separately either by block and tackle or manually. This arrangement represents a reasonable technical solution. It is by no means certain that it is the best scheme from a cost standpoint. To satisfy himself on this score the designer, unless he is very experienced, would have to try several arrangements and compare them on an overall cost basis.

2.5.2   Example 2

Design of full height formwork for a reinforced concrete abutment wall. Abutment wall is 800 mm thick, 7.5 m high, in bays 9.5 m long.

Data: Maximum concrete pressure = 80 kN/m²

Design brief: Use birch-faced plywood facing, sawn timber walings, mild steel channel soldiers and proprietary heavy duty tie rods, working load 100 kN.

Falsework to be assembled in large panels for lifting by crane.

Concrete pressure diagram
Depth to maximum pressure $= 80/25 = 3.2$ m.

Tie rod spacing
Select pitch of soldiers $= 750$ mm

$$\text{Tie spacing} = \frac{\text{tie load}}{\text{concrete pressure} \times \text{solider spacing}} \times \frac{1}{1.17}$$

$$= \frac{100}{80 \times 0.75} \times \frac{1}{1.17} = 1.425 \text{ m}$$

Effect of continuity in soldiers will increase tie rod loading, spacing reduced by 17%.

Solider loading diagram (fig. 2.13)
Make AB $= 150$ mm, position F to take account of reduced pressure and eliminate one tie rod position.

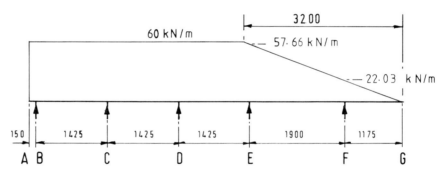

Figure 2.13 Soldier loading diagram

Analysis by moment distribution
Refer to *Steel Designers' Manual* or other textbook for a fuller explanation of the method.[5]

The beam (soldier) is free to rotate at B and F and is fixed in a horizontal position at C, D and E. CB and EF may be regarded as propped cantilevers when calculating fixed end moments (fig. 2.13). If the span stiffness factor is K the stiffness factor for these spans is $\frac{3}{4}K$.

$$\tfrac{3}{4}K_{\text{BC}} = \frac{0.75}{1.425} = 0.526, \quad K_{\text{CD}} = \frac{1}{1.425} = 0.702,$$

$$K_{\text{DE}} = \frac{1}{1.425} = 0.702, \quad \tfrac{3}{4}K_{\text{EF}} = \frac{0.75}{1.9} = 0.395$$

The distribution factors (D.F.) are given by the following equations:

$$\text{D.F.}_{\text{CB}} = \frac{\tfrac{3}{4}K_{\text{BC}}}{\tfrac{3}{4}K_{\text{BC}} + K_{\text{CD}}} = 0.428, \quad \text{D.F.}_{\text{CD}} = 1 - \text{D.F.}_{\text{CB}} = 0.572$$

$$\text{D.F.}_{\text{DC}} = \frac{K_{\text{CD}}}{K_{\text{CD}} + K_{\text{DE}}} = 0.5, \quad \text{D.F.}_{\text{DE}} = 0.5$$

Table 2.3 Distribution table – moments in kNmm

	A	B	C	D	E	F	G
Distribution factor			0.428 0.572	0.5 0.5	0.64 0.36		
Cantilever moment	+675	−675			+5069	−5069	
Carry over (C.O.)		−337					
Fixed end moment		+15 230	−10153 +10153	−10153 +10153	+2535 −18516		
Distribution		−2029	−2711	+3730	+2098		
C.O.			−1356	+1865			
Distribution			−254	−254			
C.O.			−177	−177	+113		
Distribution			+76 +101	+113	+64		
C.O.				+50 +56			
Distribution			−53	−53			
C.O.			−26	−26	+17		
Distribution			+9 +15	+17	+9		
C.O.				+7 +8			
Distribution			−3 +2	−7 −8			
C.O.				−4	+3		
Distribution			+1	+1	+1		
Final moments (kNmm)	+675	−675	+12950 −12950	+8540 −8540	+13809 −13809	+5069 −5069	
Reaction from end moments (kN)		−8.61	+8.61	+3.09	−3.09 +3.7	+4.6 −4.6	
Reaction from loads (kN)	9.0	42.75	42.75	42.75	42.75 42.75	43.5 32.27	12.94
Total reaction (kN)	9.0	34.14	51.36	45.84	39.66 46.45	48.1 27.67	12.94

$$\text{D.F.}_{\text{ED}} = \frac{K_{\text{DE}}}{K_{\text{DE}} + \frac{3}{4}K_{\text{EF}}} = 0.64, \quad \text{D.F.}_{\text{EF}} = 0.36$$

See fig. 2.14 for the fixed end moment values used in the moment distribution calculation in table 2.3.

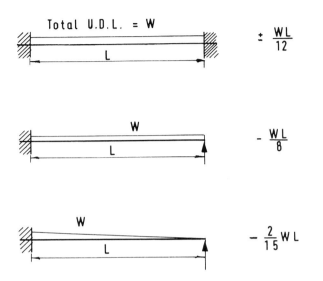

Figure 2.14 Fixed end moments

Bending moment diagram (B.M.D.)
The bending moment diagram is shown in fig. 2.15.

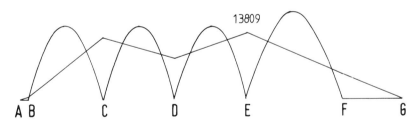

Figure 2.15 Bending moment diagram (kNmm) – maximum value noted

Shear force diagram (S.F.D.)
The shear force diagram is shown in fig. 2.16.

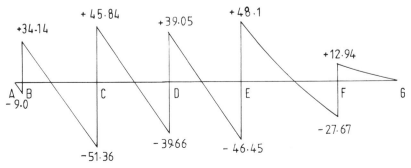

Figure 2.16 Shear force diagram (kN)

Check tie bolts
From S.F.D. maximum tie load at C = 51.36 + 45.84 = 97.2 kN which is satisfactory.

Design soldiers
From B.M.D. maximum moment at E = 13 809 kNmm
For Grade 43 steel, f = 165 N/mm²

$$\text{Required elastic modulus for single soldier} = \frac{13\,809}{165} \times \frac{1}{2}$$
$$= 41.85 \text{ cm}^3$$

From tables F.27 and F.28 127 × 64 × 14.9 kg/m Channel, with one flange bolt hole for fixing, is suitable.

Check web capacity for (i) buckling, (ii) direct bearing and (iii) shear using table F.28. Twin sections are satisfactory.

Deflections
Span EF – see fig. 2.17 for loading.

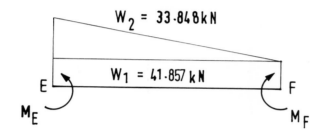

Figure 2.17 Span EF – loading

E = 210 kN/mm²
I = 965.5 cm⁴ (twin)

(i) $Md_{max} = \dfrac{1}{16} \times \dfrac{L^2}{EI} (M_E + M_F)$ N.B. upward

(ii) $W_1 d_{max} = \dfrac{5}{384} \times \dfrac{WL^3}{EI}$

(iii) $W_2 d_{max} = \dfrac{0.01304 WL^3}{EI}$

Position of maximum deflection only occurs at mid-span in condition (ii) loading. However no great loss in accuracy will result if combined maximum deflection is assumed to be at mid-span.

(i) $d_{max} = \dfrac{1900^2 (13\,809 + 5069)}{16 \times 210 \times 965.5 \times 10^4}$ = −2.1 mm

(ii) $d_{max} = \dfrac{5}{384} \times \dfrac{41.857 \times 1900^3}{210 \times 965.5 \times 10^4}$ = 1.84 mm

(iii) $d_{max} = \dfrac{0.01304 \times 33.848 \times 1900^3}{210 \times 965.5 \times 10^4}$ = 1.49 mm

Net deflection   = 1.23 mm

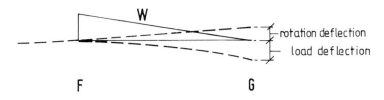

Figure 2.18 Cantilever FG – deflections

Span FG – see fig. 2.18.

Deflection due to cantilever load $W = \dfrac{WL^3}{15EI}$

$$= \frac{12.94 \times 1175^3}{15 \times 210 \times 965.5 \times 10^4} = 0.69 \, \text{mm}$$

This deflection does not take account of any rotation at F but since the deflections at the end of the cantilever due to loads on EF and FG tend to cancel out the net deflection will not be significant. The deflection in the longest span, EF, is also not significant.

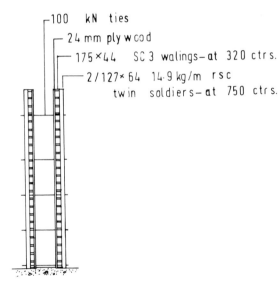

100 kN ties

24 mm plywood

175×44 SC 3 walings–at 320 ctrs.

2/127×64 14.9 kg/m rsc
twin soldiers–at 750 ctrs.

Figure 2.19 Wall formwork arrangement – full height in single lift.

Design walings and sheathing

Finnish plywood, from fig. A.4, use 24 mm with supports at 320 mm. Deflection less than 1/375.

Walings

Spacing is less than 600 mm and sheathing is continuous over at least three supports. Load sharing provisions are applicable. Bending stresses may be increased by a factor of 1.1. If continuity is allowed for the safe load may be increased by a factor of 1.25.

$$\therefore \frac{\text{Safe load with continuity and load sharing}}{\text{Safe load, simply supported, no sharing}} = \frac{1.25 \times 1.1}{1} = \frac{1.375}{1}$$

Load on waling $= 80 \times 0.32 \times 0.75 = 19.2\,\text{kN}$

Equivalent load W in basic graphs $= \dfrac{19.2}{1.375} = 14.4\,\text{kN}$

From load/span graph 2 (fig. A.3), 175×44 strength class SC3 is satisfactory. Deflection is not a critical criteria but to check it use

(a) $E = 7344\,\text{N/mm}^2$, mean value in load sharing condition.

(b) Deflection $= \dfrac{1}{145} \times \dfrac{WL^3}{EI}$, maximum deflection of continuous beam.

$$= \dfrac{1}{145} \times \dfrac{19.2 \times 10^3 \times 750^3}{7344 \times 19.7 \times 10^6} = 0.39\,\text{mm}.$$

Proprietary soldiers

The bending moments and loads in this example are within the capability of many of the soldiers described in table 2.1. Their use would normally be considered as an alternative at this stage.

Plate 28 Set of full height mushroom-head column forms on travelling gantry

Plate 29 Adjustable steel forms for piers of varying cross-section

Plate 30 Full height pier form with yokes to eliminate through ties

Plate 31 Cantilever formwork on buttress dam construction

Plate 32 Chimney wind shield construction

Plate 33 Room form. Steel faced formwork with soffit and walls in one piece

Plate 34 Wall construction with timber forms and steel soldiers

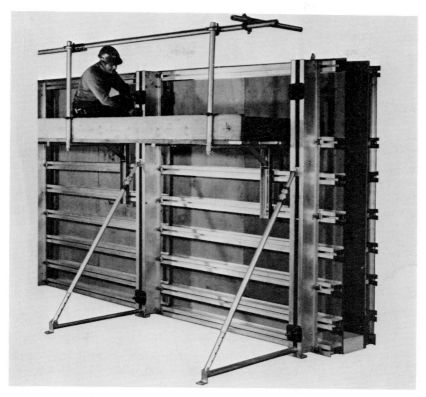

Plate 35 Aluminium walings and soldiers in wall formwork

Plate 36 Steel octagonal column forms

Chapter 3
Deck and Floor Falsework

Reinforced concrete slab construction falls into two main categories, (i) floor slabs in multi-storey buildings and (ii) bridge decks. What differences there are in construction methods and materials arise from the nature of the permanent works. Falsework loads in building works are generally light with propping heights less than 4 m. The repeating arrangements, particularly if well designed on modular dimensions, make it economic to use proprietary systems with decking and centering designed as an integrated unit. Very rapid striking and re-erection is made possible by the use of semi-permanent props which are left in place, perhaps over several successive floor levels, enabling nearly all of the decking material to be re-used immediately. Bridge decks are much heavier with greater propping heights. Most bridge falsework arrangements will involve a 'one off' design. The complexity will vary with the nature and location of the permanent works. This chapter deals with falsework arrangements and materials for slabs generally. The method used in the design example may be applied to all types of slabs. More elaborate types of bridge falsework are introduced in following chapters. Special features of design are dealt with in worked examples.

3.1 FLOOR SLABS

Drawings and photographs in this section have been selected to illustrate typical falsework arrangements for various types of floors.

Beam and slab floors
Usually beams and slab are cast at the same time, Plate 37. The increasing popularity of structural steelwork for high rise frames and the use of precast flooring and roofing units has resulted in a reduction in this type of construction. However, it still occurs frequently and suppliers provide a range of solutions.

In steel framed buildings the existing beams can be used to support the decking for an *in situ* floor. If the beam is to be cased, either for fire protection or appearance, the beam soffit shutter can be suspended from m.s. hangers over the beam and used in turn to support beam sides and deck formwork (fig. 3.1). This arrangement appears very efficient as it eliminates propping, however unless the height is large it may be more economic to support from the previous level. Expanded metal serves as combined shuttering and reinforcement for a multiple arch floor (fig. 3.2).

Beams are not always part of an *in situ* floor. Many types of building involve the construction of independent beams. These may be very deep if a large clear span is required or if the beam forms part of a heavy plant foundation.

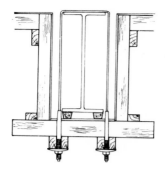

Figure 3.1 Hangers on steelwork to support formwork to beam casing

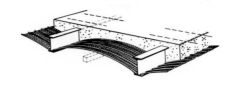

Figure 3.2 Expanded metal as permanent soffit shuttering and reinforcement

Flat slabs

Most reinforced concrete designers now recognise that to reduce the volume of concrete in a floor by having beams or local thickening at columns is counter productive in cost terms. Additional expense of formwork will more than eliminate any notional saving in concrete. Flat slabs, even with irregular or circular column heads, are an ideal application for a decking system.

Trough and waffle floors

Plate 38 shows the finished appearance of a floor constructed using polypropylene waffle units.

Standard beam arrangements

BS 5975 provides standard solutions for beam support arrangements for a useful range of beam sizes up to 1000 mm deep. These standard solutions are applicable only if the loading criteria and standards of materials and workmanship outlined in BS 5975 are maintained.

3.2   FORMWORK MATERIALS AND COMPONENTS

Most of the materials used have been described in the previous chapter. Systems based on face panels are dual purpose and are used for walls and soffits. Aluminium sections are very versatile and intended for both wall and floor construction (Plate 40). Dressed timber boarding, with a greater rigidity than plywood, is used for beam soffits and sides. Also for staircase risers and stringers.

Beam struts

Figure 3.3 illustrates a proprietary adjustable beam strut.

Beam clamps

Figure 3.4 shows the use of proprietary beam clamps. In some assemblies most of the formwork can be removed for re-use leaving a semi-permanent prop until the concrete has gained sufficient strength.

Adjustable floor centres

Floor centres are made in a variety of sizes to deal with a range of loadings and spans. They are made of steel in lattice or box construction and are telescopic to provide

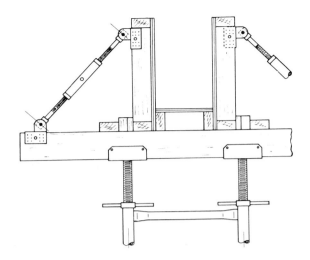

Figure 3.3 Adjustable beam strut used with timber formwork

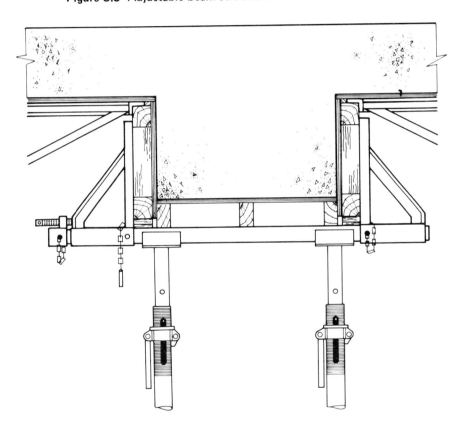

Figure 3.4 Adjustable beam clamp with removeable end section

length adjustment and easy stripping. Floor centres used without an intermediate support require to be set to a predetermined camber as they deflect significantly under load. Plate 41 shows floor centres in use with typical support arrangements.

Decking formwork and centering

Figure 3.5 illustrates 'traditional' slab falsework. Manufactured products can be substituted for part or all of the component parts. Table 3.1 lists some of these alternatives.

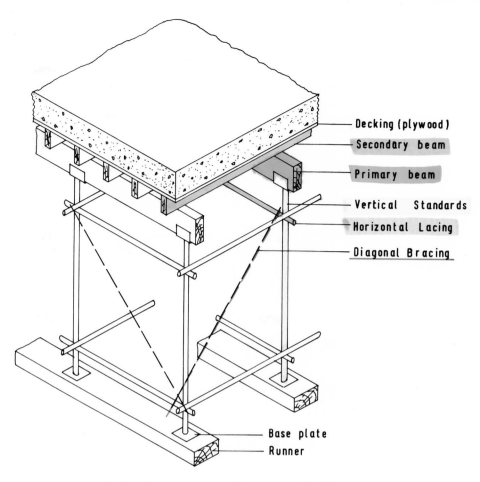

Decking (plywood)
Secondary beam
Primary beam
Vertical Standards
Horizontal Lacing
Diagonal Bracing
Base plate
Runner

Figure 3.5 Conventional falsework with decking and scaffolding

Table 3.1 Alternative formwork and support materials

'Traditional'	Alternatives
Plywood sheeting	(i) Steel framed panels with steel or plywood facing (ii) Aluminium framed panels with plywood facing (iii) Trough or waffle units (iv) Permanent soffit shuttering wood wool slabs asbestos sheeting prestressed concrete planks expanded metal profiled sheet steel
Timber primary and secondary beams	(i) Mild steel sections (ii) Aluminium sections (iii) Lattice system joists (iv) Adjustable floor centres
Scaffold tube and fittings	(i) Adjustable steel props (ii) Proprietary framed support systems

3.3 MODULAR DECKING SYSTEMS

A feature of the three proprietary decking systems illustrated in figs. 3.6 – 3.8 is an 'early strip' capability. In common with most systems the use of specially designed drop head enables the standard to remain in continuous contact with the soffit concrete, acting as a prop while the sheeting, beams, etc., are removed for re-use as soon as the concrete has gained sufficient strength.

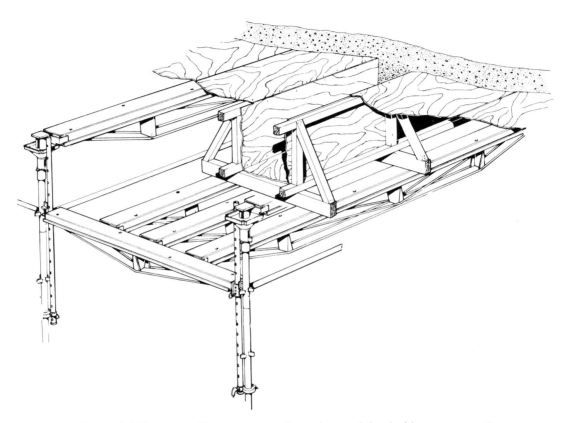

Figure 3.6 Downstand beam construction using modular decking components

3.4 SUPPORT AND CENTERING MATERIALS

3.4.1 Scaffold tube and fittings to BS 1139
Effective lengths in tube and coupler scaffolding: See fig. F.2.

Maximum permissible axial stresses and loads in steel scaffold tubes: See table F.13.

Safe working loads for individual tubes and couplers: See table F.15.

Scaffold tube properties and data: See table F.11.

Erection tolerances and workmanship: See appendix section A.6.

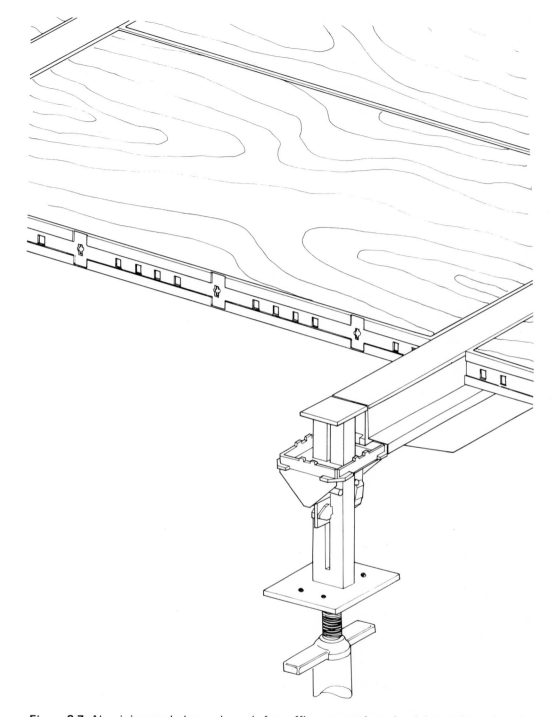

Figure 3.7 Aluminium and plywood panels for soffit supported on aluminium primary beams

3.4.2 Adjustable steel props to BS 4074[23]

(As illustrated in fig. 3.9.) Maximum and minimum propping heights are given in table 3.2.

Safe working loads: See appendix section A.5.

Erection tolerances and workmanship: See appendix section A.6.

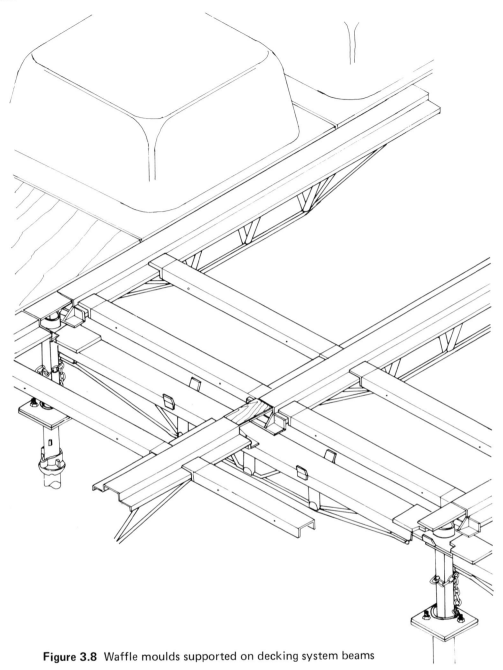

Figure 3.8 Waffle moulds supported on decking system beams

Table 3.2 Maximum and minimum propping heights

Prop size no.	Max. (mm)	Min. (mm)
0	1820	1070
1	3120	1750
2	3350	1980
3	3960	2590
4	4870	3200

Figure 3.9 Telescopic prop

3.4.3 Proprietary soffit beams – non-system

In addition to purpose-made beams which are part of a centering system, and adjustable floor centres previously described, a number of lightweight proprietary beams are available to replace conventional timber bearers. These extruded aluminium or pressed steel sections are also designed for use as walings or soldiers in wall formwork.

Table 3.3 gives a brief description of some of the items available. Manufacturer's technical literature should be consulted for full design specifications.

Table 3.3

(i) Kwikform lightbeam

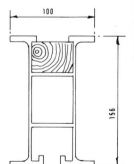

lengths: five No. 1800 – 7200
weight: 5.38 kg/m
max. permissible bending moment: 9.31 kNm
max. reactions, end: 43.6 kN
                    intermediate: 56.6 kN

Table 3.3 (*contd.*)

(ii) Kwikform extra-lightbeam

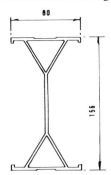

lengths: 3600, 4800 and 7200
weight: 2.75 kg/m
max. permissible bending moment: 5.42 kNm
max. reactions, end: 34.4 kN
                    intermediate: 44.7 kN

(iii) GKN aluma beam

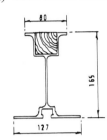

lengths: seven No. 1829 – 5486
weight: 6.0 kg/m
max. permissible bending moment: 8.3 kNm
max. permissible reaction (150 mm): 42.7 kN

(iv) RMD rapid channel

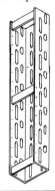

lengths: 1200, 2400 and 3600
weight: 5.83 kg/m
moment of resistance: 2.45 kNm
max. reaction: 8 kN

3.4.4 Proprietary scaffold support systems

Proprietary systems are based on either single standards or two-leg frames light enough to be erected manually. Most have common features designed to speed erection and reduce the level of skill required. Patented clip or wedge fixings requiring no bolts or special tools are used in place of loose fittings. For maximum adaptability basic components are supplied in a variety of modular sizes and have a range of purpose-made fittings. Standard scaffold tube and fittings may be used for additional lacing and bracing.

Frame systems

(i) Acrow 'Shorebrace'

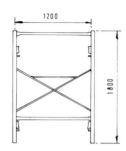

Max. safe working load $= 53.4$ kN/leg with a factor of safety of 3

(ii) RMD 'Strongshor'

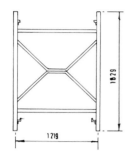

Max. load $= 45$ kN/leg

(iii) 'Millshore'

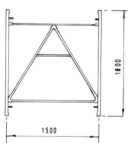

Max. load $= 59.8$ kN/leg

(iv) Mabey 'Trestlex'

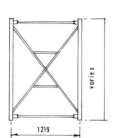

Capacity $= 52.5$ kN/leg

Standards
(i) 'Kwikstage'

Load = 39.9 kN/leg when braced at 1498 crs

(ii) SGB 'Cuplok'

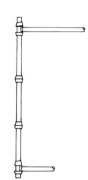

Load = 57 kN/leg laced at 1000 crs

3.4.5 Heavy duty shores and towers

For support arrangements which require individual props to take high loads a variety of fabricated sectional lattice shores are available. They can be fitted with alternative end assemblies and may be used also as raking or horizontal props.

(i) RMD 'Trishore'

Light (illustrated): Max. vertical load = 200 kN for 5.11 m height
Heavy: Max. vertical load = 3000 kN for 11.5 m height

(ii) 'Milltress 20'   Safe working load = 199 kN for 6 m between bracing

(iii) Acrow high duty   Safe working load = 81 kN for 5 m height

(iv) Mabey trestling

Normal minimum working load = 450 kN/leg

3.4.6 Heavy duty bridging and arch girders
(i) Mabey 'Universal' bridging

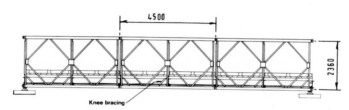

'Universal' bridging is a development of Bailey bridging of military fame. Although designed as a permanent bridge for standard road traffic it is versatile enough to be valuable in a variety of temporary works including bridge falsework. See Plate 43.

(ii) RMD 500

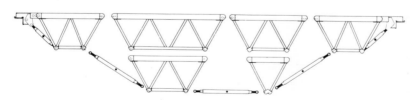

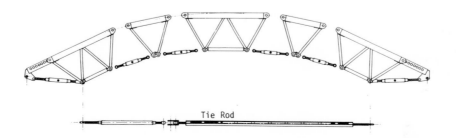

The basic component is a lightweight lattice girder which can be assembled into a wide range of spans. The profile geometry may be varied as shown to give a slight camber or full arch.

(iii) RMD H33

Heavy duty lattice-type girder, overall height 2.14 m, see Plate 44.

Plates 42, 45, 46, 47 and 51 illustrate typical support falsework problems solved by a variety of methods.

3.5 DESIGN EXAMPLE USING SIMPLIFIED METHOD FOR A LOAD-BEARING BIRDCAGE SCAFFOLD

3.5.1 Introduction to method

Any falsework structure should be designed so that at all stages of construction, under any possible combination of loading, factors of safety will be maintained with regard to:
  (a) structural strength of individual elements and connections;
  (b) lateral stability of individual elements and the structure as a whole;
  (c) overall stability against overturning and sliding.

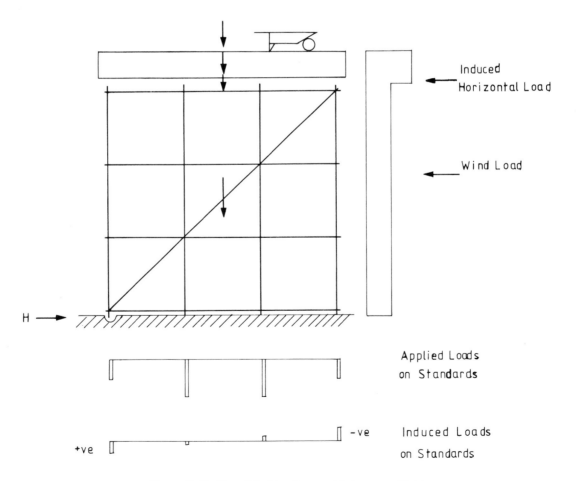

Figure 3.10 Simplified loading on birdcage scaffold

Failure of the load-bearing birdcage scaffold in fig. 3.10 will occur if any of these design conditions is not met. In practice this may be simplified to:

(a) overload of standards under worst combination of applied vertical load and induced load;
(b) failure of standards under sway produced by combination of vertical and horizontal loads, i.e. inadequate diagonal bracing;
(c) overturning. The danger time for the falsework blowing over will usually occur when the shuttering is complete before the reinforcement is placed. Other stages of construction, including the maximum wind force after concreting, may have to be checked.

A full and elegant analysis of even the simple structure in fig. 3.10 would be extremely difficult because of the problematic nature of fixity at joints. This method does not attempt to solve this intractable problem but instead creates a simplified model of how the structure will behave by making certain design assumptions:

(a) space frame action of the birdcage scaffold may be ignored and loaded members designed as individual elements, except for overall stability when it is assumed to act as a rigid box;
(b) scaffold fittings impart no fixity and no moments are transferred or induced at node points;
(c) vertical standards take all vertical loading to footings;
(d) horizontal lacers are not loaded but serve to reduce the effective length of standards;
(e) diagonal bracing takes all horizontal loading down to fixed footing.

3.5.2 Example
Design suitable deck falsework for the footbridge shown in fig. 3.11 using timber formwork and scaffold centering.

3.5.3 Design procedure
Note that design decisions are entered in the calculations with minimum explanation. The prior considerations which resulted in this data being used are discussed in section 3.5.4.

(a) *Loading*

Self-weights   (i) scaffolding = 0.5 kN/m² $\}\ \varepsilon = 1$ *
               (ii) formwork = 0.5 kN/m²

N.B. check when final arrangement is known.

Imposed loads   (i) reinforced concrete deck at 2500 kg/m³ = 11.25 kN/m²
               (ii) construction operations   = 1.5 kN/m²

Environmental loads – see 'Wind loading' (d).

Total vertical load on standards is 11.25 + 1.5 + 1.0 * = 13.75 kN/m² ✓

(b) *Layout of standards*

Primary beams are 225 × 75 class SC3 timber at 1.25 m centres. (Trial dimensions which appear reasonable are selected at this stage.)

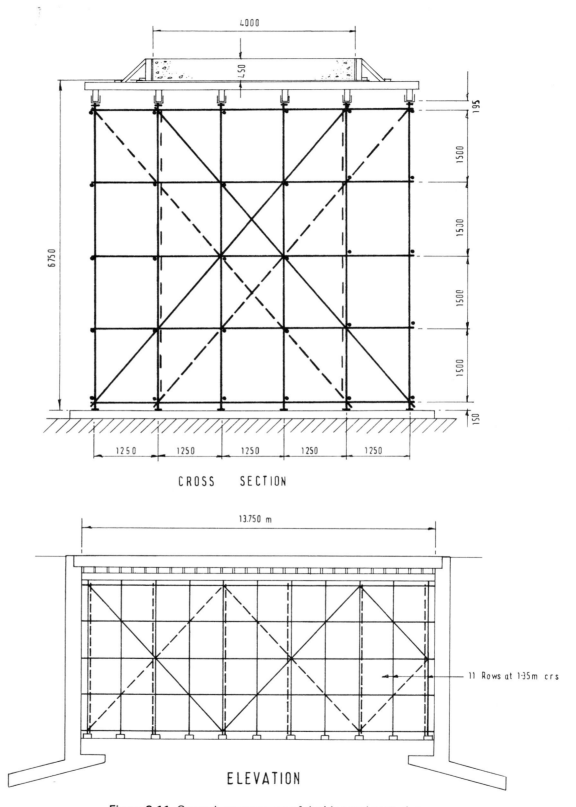

CROSS SECTION

ELEVATION

Figure 3.11 General arrangement of decking and centering

Loading on primary beam = $(11.25 + 1.5 + 0.5) \times 1.25 = 16.56\,\text{kN/m}$

page 130

from graph 5 (fig. A.3) safe span   = 1.35 m

use 1.35 (longitudinal) $\times$ 1.25 (transverse) grid

16.56 kN/m.

1.35 m.

(c) Lacing

To allow for the effect of continuity of the primary beam on support reactions increase the applied vertical load by 10%. To allow for the vertical loads induced by horizontal loads increase the applied vertical loads by, say, 2%.

Estimate of total vertical load on standards = $13.75 \times 1.12 = 15.4\,\text{kN/m}^2$

Design load per standard = $15.4 \times 1.35 \times 1.25$ = 25.99 kN

From table F.13 maximum effective length for used tubes = 1900 mm.

Figure 3.11 illustrates a suitable arrangement. At this stage the self-weight of scaffold should be checked.

(d) Wind loading

For a sheltered site in Edinburgh

falsework height	= 6.75 m
basic wind speed, V	= 50 m/s
topography factor, S_1	= 1
surface condition and height, S_2	= 0.65
statistical factor (probability), S_3	= 0.77
Design wind speed $V_s = VS_1S_2S_3$	= 25 m/s
Dynamic wind pressure, q, from table E.1	= 383 N/m²

$$\text{Maximum wind force (N/m)}, \quad W_m = qA_e\,C_f\eta$$

where A_e is the effective frontal area

C_f is a force coefficient related to shape (table E.2 appendix E)

η is shielding factor, taken as 1 for scaffolding.

Scaffolding.

Tube area averages 8% of total, for six rows 6345 mm high, take C_f as 1.3.

$$\text{Max. wind force on scaffolding}, W_m = 383 \times 6 \times 0.08 \times 1.3 \times 6.345$$
$$= 1.52\,\text{kN/m}$$

Falsework.

(i) soffit and beams: $C_f = 2.0$, $\eta = 1.0$, $A_e = 0.405\,\text{m}^2/\text{m}$ (fig. 3.11).

$$\text{Max. wind force} = 383 \times 0.405 \times 2 \times 1 = 310\,\text{N/m}$$

(ii) edge formwork: $C_f = 1.8$, $\eta = 1.0$, $A_e = 0.450\,\text{m}^2/\text{m}$.

$$\text{Max. wind force (windward)} = 383 \times 0.45 \times 1.8 \times 1 = 310\,\text{N/m}$$

(N.B. Wind force on the leeward edge formwork is the same and will apply when the formwork is erected ready for concrete.)

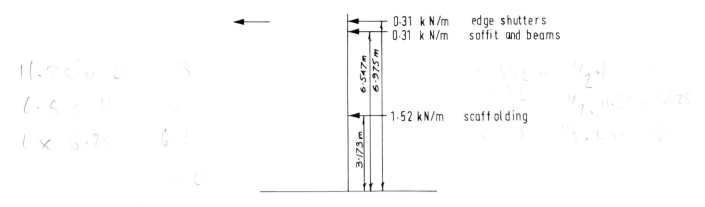

Figure 3.12 Wind load diagram

From the wind force diagram, fig. 3.12

Total maximum wind force, W_m $= 1.52 + 0.31 + 0.31 = 2.14\,\text{kN/m}$
Total moment of maximum wind force, M_{W_m} $= 4.82 + 2.03 + 2.16 = 9.01\,\text{kNm/m}$

(e) Distribution of applied vertical loads
These are illustrated in fig. 3.13.

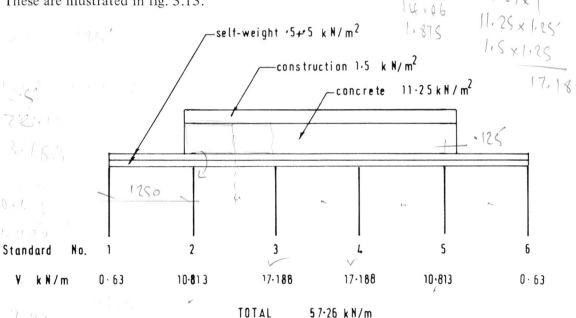

Figure 3.13 Applied vertical loads

(f) Horizontal forces
(i) Horizontal force equivalent to 2.5% of applied vertical loads, H_v is 2.5% of $57.26 = 1.432\,\text{kN/m}$ (applied at soffit level).

$$\text{Moment, } M_{H_v} = 1.432 \times 6.75 = 9.67\,\text{kNm/m}$$

(ii) Horizontal force resulting from erection tolerances, 1.0% of applied vertical load, H_t is 1% of $57.26 = 0.57\,\text{kN/m}$ (applied at soffit level).

$$\text{Moment, } M_{H_t} = 0.57 \times 6.75 = 3.85\,\text{kNm/m}$$

(iii) Wind forces from (d).

$$\text{Max. wind moment, } M_{W_m} = 9.01 \text{ kN/m}$$

(Use greater of (i) or (ii) or [(ii) + (iii)]/1.25.)

(g) *Combine applied and induced vertical loads and compare actual and estimated maximum loads on standards*

Induced vertical forces.

 (i) forces arising from applied moments may be positive or negative.
(ii) in table 3.4 loadings are expressed in kN/m for as long as possible to facilitate recalculation.

Table 3.4 Combined applied and induced loading

		1 & 6	2 & 5	3 & 4
Standard reference (from fig. 3.13)		1 & 6	2 & 5	3 & 4
Proportion of applied moment, p		0.357	0.129	0.014
Lever arm, l (m)		3.125	1.875	0.625
Induced vertical loads	V_{H_v}	1.160	0.665	0.217
Example:	V_{H_t}	0.440	0.265	0.086
V_{H_v} = vertical load arising				
from moment M_{H_v}	V_{W_m}	1.029	0.620	0.202
$= \dfrac{p M_{H_v}}{l}$				
Applied vertical loads (from fig. 3.13)		0.63	10.813	17.188
Combined vertical loads (max.) $V + V_{H_v}$		1.790	11.478	17.405
$V + V_{H_t}$		1.070	11.078	17.274
$\dfrac{V + V_{H_t} + V_{W_m}}{1.25}$		1.679	9.358	13.981

$$\text{Max. load per standard} = 17.405 \times 1.35 = 23.497 \text{ kN}$$
$$\text{Add 10\% for continuity overload} = 2.350$$
$$\text{Total} = 25.847 \text{ kN}$$

$$\text{Estimated max. load per standard} = 25.99 \text{ kN}$$

Trial arrangement is satisfactory

(h) *Diagonal bracing*

Design horizontal loading/row at 1.35 m centres.

$$\text{Use maximum of } H_v = 1.43 \times 1.35 = 1.93 \text{ kN}$$
$$H_t = 0.57 \times 1.35 = 0.77 \text{ kN}$$
$$\frac{H_t + W_m}{1.25} = \frac{(0.57 + 2.14) \times 1.35}{1.25} = 2.93 \text{ kN}$$

N.B. Wind loading not applicable to longitudinal bracing.

Design criteria for diagonal bracing.

Use least of (i) coupler capacity of 6.25 kN
(ii) safe load of diagonal as a strut.

(The latter is not likely to be critical in this example as diagonals will be fixed at each level of lacing.)

transverse

θ = angle to horizontal = $\tan^{-1}\dfrac{1.5}{1.25}$ = 50.194°.

Using (i), number of rows/diagonal

$$= \frac{6.25 \cos\theta}{2.93} = 1.36$$

Use double brace every second row for symmetry.
Check (ii), load on brace

$$= \frac{2.93}{\cos\theta} = 4.57\,\text{kN}$$

From table F.13 this is more than satisfactory.

longitudinal

$$\theta = \tan^{-1}\frac{1.5}{1.35} = 48.01°$$

$$\text{Total horizontal load} = 1.93 \times 10\,\text{kN}$$

$$\text{Number of diagonals} = \frac{1.93 \times 10}{6.25 \cos\theta} = 4.62 \ (5 \text{ used in arrangement})$$

(*j*) *Overall stability*
Falsework should normally be designed to withstand all reasonable combinations of vertical and horizontal loads. In all cases the factor of safety against overturning should be at least 1.2. In the example maximum wind force acting on the completed formwork without any stabilising imposed loads such as reinforcement or plant will be the critical overturning condition.

$$
\begin{aligned}
\text{Overturning moment previous} &= 9.01\,\text{kNm/m} \\
\text{add leeward} &= 2.16\,\text{kNm/m} \\
\text{add } H_t \text{ component} &= 0.42\,\text{kNm/m} \\
\hline
\text{Total} &= 11.59\,\text{kNm/m}
\end{aligned}
$$

Righting moment from self-weight of formwork and scaffolding

$$= (0.5 + 0.5) \times 6.25 \times 3.125 = 19.53\,\text{kNm/m}$$

$$\text{Factor of safety} = \frac{19.53}{11.59} = 1.69$$

No kentledge or other form of holding down required.

(*k*) *Foundations*
No reference has been made so far to foundations. Normally the first step in any falsework calculation where ground bearing, either on fill or original ground, is a

consideration would be a careful assessment of ground conditions and bearing capacity. If necessary trial loadings should be made. This is particularly so when it is proposed to concentrate vertical loads into towers. The viability of the arrangement may be totally dependent on ground conditions so it is important to establish suitability at an early stage.

In this example examination of borehole data, confirmed by inspection during abutment excavations, yields a bearing pressure of $250 \, kN/m^2$ for undisturbed ground. Adjacent to abutments selected back-fill will be placed and compacted to give a safe bearing pressure of $150 \, kN/m^2$. Standards will have base plates and the load will be spread to the foundations through sole plates consisting of crossing sleepers, 300×150, in long lengths.

$$\text{Maximum load/standard} = 23.49 \, kN$$

$$\text{Maximum bearing pressure} = \frac{23.49}{0.75 \times 0.3} = 104.4 \, kN/m^2$$

This is satisfactory. See BS 5975, clause 44.4 for additional information.

(l) Important factors related to the design

These do not appear in the calculations but must be considered by the designer at this stage.

(i) Erection tolerances.

Before using safe loads in the calculations it has been assumed that scaffold tube and fittings to BS 1139 will be used and that all material will be in good condition. Also that the standard of workmanship, particularly with respect to erection tolerances will be in accordance with BS 5975. The relevant information should be clearly stated on the drawings and a system of site inspection and checking carried out.

(ii) Initial camber

Plates 49 and 50 illustrate graphically the need to ensure that the final soffit profile is within acceptable limits. This will only happen if positive steps are taken to make sure that it is.

All possible areas of settlement and deflection must be examined and an allowance made, based on the best information. Ground settlement is best measured directly, on most roadworks contracts suitable mobile testing plant will be available. Foundations should be monitored continuously to check for deterioration. Proper drainage may be required. Some types of fill are very susceptible to frost heave. Elastic shortening of standards can be easily calculated. (In the example maximum shortening is 1.28 mm.)

There will be some 'take up' at points of contact when reinforcement is placed but further 'give' can be expected when concreting starts. Provided all bearing points, forkheads, and base plates have been inspected and tightened a reasonable allowance for this would be 5 mm where timber beams are used. The bridge will, of course, deflect under its own weight and an allowance for this should be made. Since any sag will be more apparent than even a large amount of hogging it is always better to allow a generous initial camber when setting out the soffit levels.

3.5.4 Design procedure – discussion and explanation

(a) Loading
Self-weights.

The total self-weight should include:

(i)   the falsework structure
(ii) any ancillary temporary works connected to the falsework e.g.
    (1) access ramps
    (2) hoist or other tower structure
    (3) loading storage platforms
    (4) raking and flying shores
(iii) the formwork
(iv) any permanent works elements forming an integral part of the falsework.

In the example (i) and (iii) are relevant. The initial estimates of self-weights should be checked at the appropriate stage in the calculations.

Imposed loads.

Imposed loads should include the loading from:

(i)   permanent works, e.g. reinforcement and concrete
(ii) construction operations including:
    (1) working areas
    (2) storage areas
    (3) pedestrian traffic
    (4) vehicular traffic
    (5) static plant
    (6) mobile plant.

Figure 3.11 shows how the horizontal components of concrete pressure are resisted without imposing any loads on the centering. Only the vertical dead weight of the concrete is imposed on the falsework. The secondary beam acts as a tie to take the horizontal component – appropriate notes must be added to the drawings with the instruction to bolt any lap joints in the timber. The vertical component of concrete pressure is small enough to be ignored in this case.

The loading of $1.5 \, kN/m^2$ is that recommended in BS 5975 for construction operations and allows for:

(i)   construction operatives
(ii) hand tools and small concreting equipment
(iii) materials required for immediate use
(iv) dropping and heaping of concrete to specified limits.

In the example the possibility of other imposed loads has been examined and deemed to be not relevant.

Environmental loads.

Depending on location and time of year a combination of environmental loads may have to be assessed.

(i)   wind loading
(ii) water. Forces produced either by flowing water or wave action.
(iii) snow

(iv) ice

(v) earth pressure.

In the example only wind loading is relevant.

(b) and (c) Layout of standards and Lacing

A variety of arrangements is possible at this stage in the design unless the spacing of beams has been predetermined, say by the selection of a proprietary decking system. The grid should be reasonably square, with simple spacing to fit the overall dimensions of the permanent work.

Soffit formwork.

Decking plywood; loading = 12.75 kN/m².
From fig. A.4 15 mm birch faced plywood with supports at 400 mm centres is suitable, deflection less than 1.5 mm.
Secondary beams; load = 1.25 × 0.4 × 13.25 = 6.63 kN
From fig. A.3 graph 3 150 × 50 class SC3 timber is suitable.

Design note – effect of continuity.

(i) Beams.

Primary and secondary beams are treated as simply supported with uniformly distributed load. This condition is not strictly true as beams may be continuous over several supports and the actual loading on primary beams is a series of point loads. A detailed analysis would not be justified even if exact sizes could be specified. Although an argument could be advanced for using a reduced bending moment, say $WL/10$, it should be noted that:

  (a)   random lengths of timber will be used and it is possible that a short length, equal to a span, will be placed.
  (b)   concrete load is applied progressively, at some stage a single span will be loaded.

Shear in beams at supports will be underestimated if continuity is not considered, this is unlikely to be significant.

(ii) Standards.

When loads on standards are calculated without considering the effect of continuity BS 5975 recommends that the loading is increased by 10% to allow for variations of the transferred loads. This value has been used in the example.

When random lengths of primary and secondary timber are used, and placed with joints systematically staggered, then an additional 10% is a reasonable allowance for reactions at standards. It should, however, be regarded as a minimum value. In fig. 3.14 the load on the centre standard is 56% greater than that calculated by assuming simply supported conditions for all spans of secondary and primary beams.

Arrangements similar to fig. 3.14 are most liable to occur when proprietary joists of equal length are being used. In this case the 10% increase is not appropriate, a more detailed analysis will be required.

(d) Wind loading

The procedure for calculating wind loading on falsework described in BS 5975 has

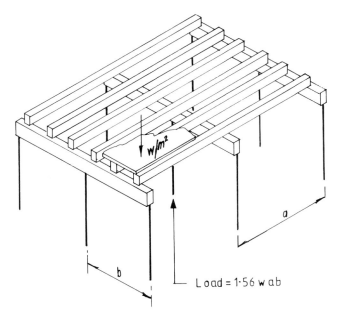

Figure 3.14 Effect of beam continuity

been followed in the example. Only maximum wind force has been calculated as this can occur during the period of greatest risk, when the deck shuttering is erected but not loaded.

Although not applicable in this example it is possible to envisage falsework arrangements which would be at risk during working operations. Lifting large units for example, or moving large areas of shuttering. The generally accepted maximum wind speed for normal working is 18 m/s which gives a dynamic wind pressure of 200 N/m². Using this value the overall stability may then be checked using the procedure in this example.

(e) Distribution of applied vertical loads
As discussed in (b) and (c) the continuity of secondary beams is neglected.

(f) Horizontal forces
The following may give rise to horizontal forces acting on falsework structures:

(1) wind forces
(2) forces resulting from erection tolerances. When an out of plumb nominally vertical member is loaded a horizontal force component is created. This may be taken as 1% of applied vertical forces acting at soffit level.
(3) forces resulting from members out-of-plumb by design
(4) concrete pressure
(5) water and wave forces
(6) dynamic and impact forces
(7) forces from the permanent structure.

To ensure the lateral stability of falsework structures they should be designed to resist at each phase of construction the applied vertical loads plus the greater of:

(i) horizontal forces equivalent to 2.5% of the applied vertical loads considered as acting at the points of contact between the vertical loads and the supporting falsework, i.e. at soffit level in the example.

or (ii) the greatest summation of (1) to (7) above.

In the example only the horizontal forces from (1) and (2) are relevant. Although it could be determined by inspection that (i) is the critical alternative in the example the full procedure has been carried out for the sake of clarity.

(g) Induced vertical loads

(i) Division of applied moment between standards.

Example – proportion of applied moment, M, to standard $1 = pM$ where $p = l_1^2/\Sigma l^2$ and l_1 is lever arm of standard 1.

$$\text{Force in standard } 1 \;=\; \frac{pM}{l_1}$$

(ii) Allowable stress.

BS 5975 recommends that if an increase in stress is due solely to wind forces the allowable stress may be increased by 25%, provided that the sections should be not less than those needed if the wind stresses were neglected. This provision has been applied in the example to the wind component of induced vertical loads and to the calculation of diagonal bracing to resist horizontal forces including wind.

(iii) Estimate of induced vertical loads.

In the estimate of probable maximum load per standard an addition of 2% of the applied vertical load was made to allow for vertical loads induced by horizontal loads. The value so obtained is 0.464 kN/standard.

The actual maximum induced vertical load in the critical standard 3 is 0.293 kN, so the original estimate was satisfactory. No adjustment or recalculation is necessary.

Note, however, that the 2% allowance was selected after observing that in the arrangement only the inner standards 3 and 4 are fully loaded. Since the outer standards attract more induced vertical load if they had also carried a greater proportion of imposed loads they might have become critical and 2% would have been too low. In the example the maximum positive induced vertical load is 1.556 kN on standard 1, equivalent to 6.72% of the applied vertical load.

As a general rule if the applied vertical load is distributed evenly:

$$\text{scaffold} \frac{\text{height}}{\text{width}} \geqslant 1 \quad \text{allow } 7.5\%$$

$$\text{scaffold} \frac{\text{height}}{\text{width}} \geqslant 2 \quad \text{allow } 10\%$$

(h) Diagonal bracing and lateral stability

In the design example lateral stability on the longitudinal axis is achieved without using the existing abutment to resist horizontal forces. However it is good practice to use the permanent structure to improve stability. In the example this could be done at very little additional expense by packing both ends of the lacers to provide a continuous brace between abutments, using plastic caps or timber packing to protect the concrete.

(j) Overall stability

Stability against overturning is satisfactory but what about sliding under the same conditions?

Maximum wind force   = 2.14 kN/m
Minimum vertical load, R = 6.25 kN/m (self-weight only)

Coefficient of static friction of softwood on granular fill, $\mu = 0.3$.
Limiting value of frictional force = $\mu \times R$
$\qquad\qquad\qquad\qquad\qquad\qquad$ = 1.88 kN/m

Sliding will occur. Unless the risk is acceptable positive steps will need to be taken. The sleepers can be anchored at the ends by driven pins, or set in concrete.

Standard solutions

The slab thickness of 450 mm in this example is within the scope of the Standard Falsework Solutions in BS 5975. However, the propping height of 6750 mm is greater than the maximum overall height limit so standard solutions could not be used in this case.

Plate 37 High rise *in situ* concrete construction

Plate 38 Waffle floor – finished appearance

Plate 39 Floor decking supported on system scaffolding and beams

Plate 40 Bridge decking with aluminium primary and secondary beams supported on system scaffolding

Plate 41 Adjustable floor centres in use

Plate 42 Soffit falsework for spiral footway ramp

Plate 43 Bailey bridging used for centering clear span and support grillages. Note the use of short fork-heads to profile the main decking beams and aluminium secondary beams

Plate 44 Heavy duty lattice girder used to give clear span support to bridge deck formwork

Plate 45 Bridge arch centering formed with lattice sectional girders

Plate 46 Combined soffit and falsework flying form made in aluminium

Plate 47 System decking and support scaffolding

Plate 48 Construction of suspended span. Note kentledge blocks on projecting arms of prestressed cantilever

Plate 49 Ultimate profile of suspended span on cantilevered bridge not as calculated. See section 3.5.3(ii)

Plate 50 Insufficient camber allowed for take-up and compression

Plate 51 Bridge deck soffit formwork in timber supported on system scaffolding

Chapter 4
Falsework Supervision and Procedures

Every review of falsework practice[15,16] has included comment and recommendations on procedure. The Bragg Committee[20] in particular identified weaknesses in the organisation of falsework design and construction practices. Their studies indicated that while falsework failures occur for a variety of reasons there are usually two common elements:

(a) technical error in design or construction which led to collapse;
(b) procedural inadequacies which allowed the faults to go undetected and uncorrected.

Falsework schemes vary greatly in scale of work and degree of managerial complexity. Problems of coordination in a small support scheme using a standard falsework solution will obviously be less than in a large contract with several concurrent falsework operations. This may make the establishment of a procedural control system difficult as flexibility will have to be built in to make it possible to cope with a variety of physical and contractual conditions. Clear lines of authority, responsibility and communication must be set up to ensure that falsework schemes are designed, erected and used in a safe and satisfactory manner.

Basic functions and responsibilities, standards of care in design, workmanship, and checking will be the same regardless of the scale of the work. Only management arrangements and designations will vary from job to job. In smaller works one individual may undertake responsibility for several functions. In larger works delegation of responsibility may be necessary.

The main items for which responsibility should be established are:[1]

(a) the design brief;
(b) the concept of the scheme;
(c) the design, drawing out and specification of the falsework;
(d) the adequacy of the materials used;
(e) the control of erection and dismantling on site, including maintenance;
(f) the checking of design and construction operations;
(g) the issuing of formal permission to load and dismantle the falsework.

Table 4.1 shows diagrammatically a procedural arrangement suitable for a normal tripartite civil engineering contract where the main contractor has a contractual obligation for all aspects of the temporary works. Its main feature is the devolvement of all responsibility, except in the strict legal sense, to a falsework coordinator.

Table 4.1 Falsework responsibility/activity chart

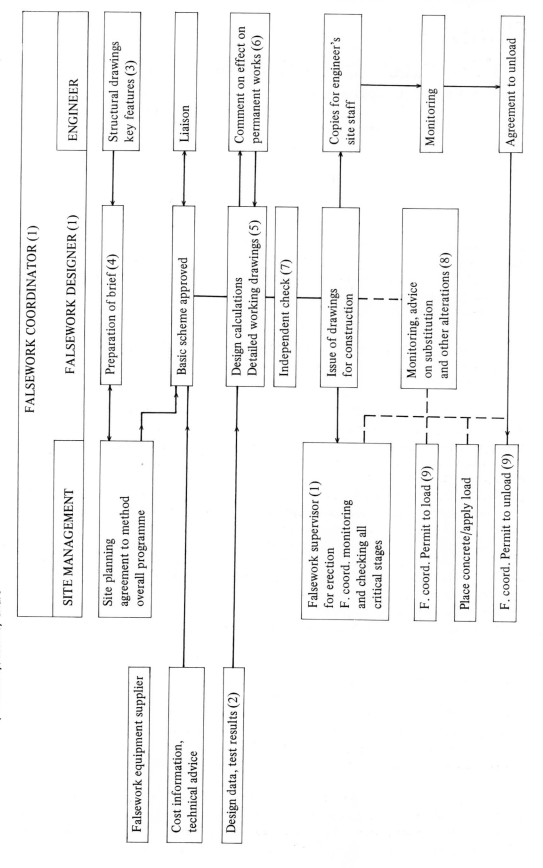

NOTES TO TABLE 4.1

(1) Job specifications

Falsework coordinator[1]
(a) coordinate all falsework activities;
(b) ensure that the various responsibilities have been allocated and accepted;
(c) ensure that a design brief has been established with full consultation, is adequate, and is in accord with the actual situation on site;
(d) ensure that a satisfactory falsework design is carried out;
(e) ensure that the design is independently checked for:
    (i) concept;
    (ii) structural adequacy;
    (iii) compliance with the brief;
(f) where appropriate, ensure that the design is made available to other interested parties, e.g. the structural designer;
(g) register or record the drawings, calculations and other relevant documents relating to the final design;
(h) ensure that those responsible for on-site supervision receive full details of the design, including any limitations associated with it;
(i) ensure that checks are made at appropriate stages covering the more critical factors;
(j) ensure that any proposed changes in materials or construction are checked against the original design and appropriate action taken;
(k) ensure that any agreed changes, or corrections of faults, are correctly carried out on site;
(l) ensure that during use all appropriate maintenance is carried out;
(m) after a final check, issue formal permission to load if this check proves satisfactory;
(n) when it has been confirmed that the permanent structure has attained adequate strength, issue formal permission to dismantle the falsework.

To ensure the independence of checks, the falsework coordinator should delegate the task to another if he himself has carried out any of the activities requiring checking.

The falsework designer[17]
His function is to design and detail falsework structures in accordance with the brief, the recommendations of BS 5975 and other appropriate documents. In particular, he should understand the detailed transfer of a load down through the formwork, if any, and through the falsework to the supporting foundation arrangements. When his own knowledge and ability are inadequate, he should seek help from appropriate experts, especially for foundation arrangements.
His duties are to:

(a) produce a correct design;
(b) provide information which can be readily understood by the people who will erect and use it.

The falsework supervisor[17]
He is the man directly in charge of the falsework construction and dismantling. He will be responsible for erecting the falsework with materials of the specified quality and conforming in detail and tolerance with the requirements of the design.

As direct line management, he is responsible for checking the falsework as it proceeds at all stages. It is very often difficult or impractical to check adequately many details after the falsework is fully erected. It is also very difficult to correct misalignment at a later stage.

When the structure is approved for loading he will be responsible to see that loading is undertaken in accordance with the agreed method.

When it is agreed that dismantling may take place, he will be responsible for seeing that this is satisfactorily carried out, and for identifying materials which have been damaged.

(2) Falsework suppliers and subcontractors

It is possible that the role of the falsework equipment supplier will not be limited to sale or hire of material and advice on its use. He may also undertake design and/or erection, under contract, for the main contractor. For certain types of falsework the main contractor may decide to use a specialist design organisation.

These alternatives do not invalidate the responsibility chain in table 4.1. Care should be taken to ensure that any additional organisations becoming involved are dovetailed into the administrative network and not allowed to operate on the periphery. Responsibility between the main and subcontractor should be carefully defined and, if necessary, written into the contractual agreement before work commences.

(3) Information from the engineer

Contractually all responsibility for temporary works devolves to the main contractor. However, it is in the interest of the work generally if the engineer participates as fully as possible when the falsework design brief is being drawn up. The following is a list of the information areas where the engineer's involvement may be required.

- (a) Full design information regarding the permanent structure, its structural action when complete and at all intermediate stages to enable the falsework designer to decide on temporary support arrangements. See Plate 48.
- (b) Full specification for construction joints, including location if dictated by design considerations.
- (c) Site investigation data including borehole logs relevant to areas of temporary works.
- (d) Environmental data collected during design stage, e.g. stage/discharge and flood/ frequency information for river works, tidal levels, etc.
- (e) Restrictions and conditions with regard to noise levels, working hours, vehicle loads, dust nuisance, etc.
- (f) Requirements of statutory authorities with regard to public access, road diversions, protection and maintenance of public utilities.
- (g) Restrictions imposed by adjacent operations, e.g. railway property, hospital.
- (h) Deflections and settlement of permanent work as it takes up self-weight loading. See Plate 49.
- (i) Special provisions for post tensioning, e.g. staging sequence, kentledge required, deflections expected, induced loads on centering.
- (k) Rescheduling of reinforcement to accommodate agreed construction plan.

(4) Preparation of brief

In addition to the information relevant to the permanent works and the engineer's requirements the design brief will include input from the main contractor's organisation.

(a) Programme of construction with emphasis on phasing to permit multiple use of falsework material.

(b) Availability of contractor's own falsework material.

(c) Costing information.

(d) Proposed site layout, contractor's access, storage, services, etc.

(e) Plant to be used, concreting arrangements.

(f) Possibility of design alternatives, e.g. precast units in lieu of *in situ.*

(5) Design calculations and drawings

Falsework design should be in accordance with BS 5975 and other relevant Standards and Codes of Practice. An important difference between design for temporary and permanent works is the application of erection tolerances in temporary works, including falsework. Working stress values in BS 5975 used in design assume that specified erection tolerances will not be exceeded. When proprietary components are used the designer should be satisfied that there is a factor of safety of at least 2 with the worst combination of loading and the worst combination of erection tolerances. These tolerances will have to be ascertained and included in the specification. Working drawings should have erection tolerances clearly indicated with emphasis on any potentially critical areas. Drawings should be site-user oriented and easy to understand. It may be appropriate to produce separate drawings when more than one trade is involved, e.g. scaffolders and formwork carpenters.

(6) Effect on permanent works

At this stage the designer of the permanent work should scrutinise the proposed construction sequence to determine its effect on his design. It should be noted that:

(a) The structural action of a partly completed structure may be significantly different to that of the finished work.

(b) Loads imposed during construction may be more onerous than the permanent design loads.

(7) Independent check

Table 6.1 describes a systematic checking procedure. It should be carried out by someone not involved in the initial design who has the seniority and experience compatible with the complexity and originality of the falsework solution undergoing the check.

(8) Monitoring by falsework designer

It is in the nature of things that site personnel will sometimes seek to make substitutions and alterations to the falsework design, usually for reasons of expediency or economy. If available beams, say, are slightly heavier than specified it is unlikely that operatives will suspect that their use may be detrimental if it significantly alters the design structural action. See fig. 6.1 for an example of this.

Figure 4.1 shows how the use of overlength beams has changed the loading conditions specified by the designer. In a complex falsework arrangement the designer is the only one competent to judge the effect of apparently minor changes. The site checking procedure should ensure that written approval is required for all substitutions and changes of detail.

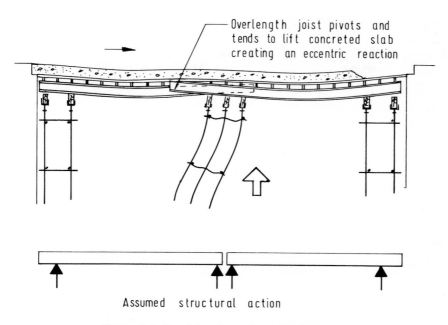

Overlength joist pivots and tends to lift concreted slab creating an eccentric reaction

Assumed structural action

Figure 4.1 Possible effect of overlength beams

(9) Permits

Checking procedure should require that written approval from the falsework coordinator is necessary before falsework is loaded or unloaded.

Chapter 5
Bridge Falsework

5.1 DESIGN METHODS

In the preceding chapters many detailed aspects of formwork design, support and safe loadings regarding the pouring of concrete to form floor and wall systems have been considered. In this section design checks for complete bridge falsework systems including overall actions resulting from self-weight, construction loads and environmental force actions are described.

For the purposes of falsework design and checking, elastic analysis and design methods are preferred over plastic or ultimate load design. This allows for close estimation of falsework deflections, stresses and stability rather than simply a check on ultimate capacity. Plastic analysis methods rely upon the determination of appropriate minimum hinge formation to provide a collapse mechanism upon which to assess the capacity of a structure. Generally falsework structures are of a braced nature and contain pin-jointed triangulated substructures either locally or globally. Such structures do not readily lend themselves to plastic analysis, the structures are often complex and unsuitable for simple plastic theory and in cases where plastic methods could be applied a design so based would require re-analysis by elastic methods to check deformation characteristics, which are all important if the intended geometry of the finished structure is to be obtained with any degree of precision. Plastic design methods and analysis require predictable joint rotation characteristics which cannot generally be guaranteed in temporary forms of structure, especially the sustenance of maximum moment capacity at the considerable rotations often required for full formation of the design mechanism.

5.2 FORCE ACTIONS

Force actions on falsework systems are many and varied. Accurate assessments are not always possible but the range of forces for everyday and infrequent occurrences or for concentrated construction activity are generally obtainable. In practice it may be that loadings prescribed by codes may act on falsework structures only on rare occasions, but generally the specified loadings are the minimum for a small probability of over-stress or failure of any structure or component and often designers use higher loading values, either globally or locally, to assess the probable performance of their designs where there is a reasonable chance of overload from one source or from a combination of forces. In other instances a designer may use his discretion to design to reduced values for certain force actions where there is a small probability of such an occurrence or if positive action can be taken to eliminate or reduce specific forces. In general terms the forces acting on falsework structures can be categorised as loadings, environmental actions and constraint effects.

Before dealing with their overall actions on bridge falsework systems it is worth commenting on a few of these force actions.

5.2.1 Main loadings

(*a*) *Dead loading* – This can be taken as the self-weight of the support structure which generally comprises a predesigned system of steel members and should therefore not be difficult to estimate to within ±20% but there is always a possibility of heavier sections being used if these are available.

(*b*) *Imposed loading* – If this is taken here to comprise the formwork, the fixed reinforcement and the poured concrete, then there is scope for considerable variation if very close control is not maintained. For instance heavier formwork than anticipated may be adopted on site and small variations in internal formwork dimensions of order ±6 mm to 12 mm from those given in plans may easily occur thus altering the quantities of poured material. Overpour or temporary heaping of concrete is also a distinct possibility.

(*c*) *Construction plant* – Direct and bending forces resulting from construction plant are often significant in the assessment of falsework structures. Although the type and positioning of such equipment may be carefully preplanned, difficulties such as breakdowns may result in the use of extra or alternative systems for lifting and placement of materials, therefore care must be taken not to underestimate force actions from such sources.

(*d*) *Storage of materials* – If not strictly controlled then severe loadings much in excess of the assumed design values may result from storage of reinforcement and secondary propping. Although temporary, such loadings may have several detrimental effects including overstress of falsework, the possibility of negative deflections on spans adjacent to spans supporting stored materials and the possibility of cracking of green concrete if storage takes place or if material is removed such that the formwork support is altered before sufficient strength has been achieved in the concrete (fig. 5.1).

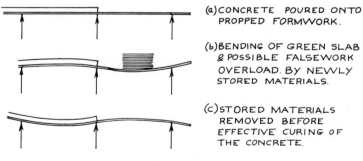

(a)CONCRETE POURED ONTO PROPPED FORMWORK.

(b)BENDING OF GREEN SLAB & POSSIBLE FALSEWORK OVERLOAD. BY NEWLY STORED MATERIALS.

(c)STORED MATERIALS REMOVED BEFORE EFFECTIVE CURING OF THE CONCRETE.

Figure 5.1

(*e*) *Impact* – Falsework design should encompass consideration of the effects of possible collisions of plant with the support structure.

(*f*) *Vibration* – Vibration may arise from the movement of plant, local transport systems, blasting, pile driving or out of balance motors for instance. Vibration should be minimised as much as possible during construction and especially when placed

Table 5.1 Guide to initiation of crack damage to green concrete caused by vibration

Peak particle velocity (mm/s)	5	50	100
Age of concrete (days)	0 - 1	1 - 7	> 7

Note – even small amplitude vibration of order 5 mm/s can cause subsidence in granular soils. Vibration above about 12.5 mm/s can initiate damage in normal housing close to a site and vibration above 50 mm/s will probably cause damage. In delicate buildings damage can occur at below 12.5 mm/s and sensitive switchgear may be tripped at around 5 mm/s.

concrete is green. A rough guide to the possibility of damage to concrete from this source is given in table 5.1. Another common effect of vibration is resonance magnification of forces in a structure where work practices or machinery cause forces in a structure at or close to one of the natural frequencies of the structure. Such action may cause more than twice the static loading on a structure from the dynamic source. The effect is further compounded by the oscillating nature of the dynamic force action which can lead to rapid fatigue of a structure or its component parts (fig. 5.2 and references 6, 12).

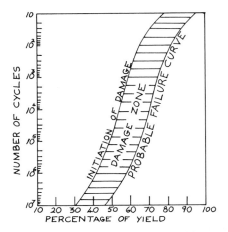

Figure 5.2 Cumulative damage from vibration, repetitive or cyclic force actions

(g)　*Other loadings* – Many other loadings such as earth pressure, fire from a burning vehicle, materials beneath a structure, debris carried by flood waters or demolished parts of the falsework may act on support structures.

5.2.2 Environmental effects

(a)　*Ground conditions* – Frost heave may affect the footings of falsework structures or footings may be caused to settle by pumping out of groundwater.

(b)　*Wind forces* – For design purposes wind is generally categorised in terms of everyday events, annual to five year return winds and extreme winds of fifty year return or greater (fig. 5.3). Exact determination of wind forces acting in a specific location, even for an exact wind speed, is extremely complex. Among the factors which affect wind action are structure aspect, orientation, wind direction, ground roughness, topography, location, the shapes and sizes of surrounding structures as well as the shape and size of the structure under consideration and the wind speed. Wind action is also affected by

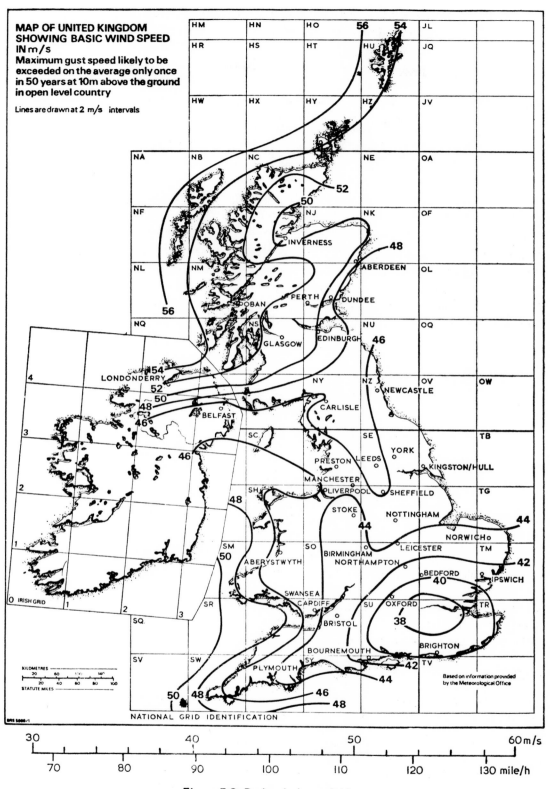

MAP OF UNITED KINGDOM SHOWING BASIC WIND SPEED IN m/s
Maximum gust speed likely to be exceeded on the average only once in 50 years at 10m above the ground in open level country

Lines are drawn at 2 m/s intervals

NATIONAL GRID IDENTIFICATION

Based on information provided by the Meteorological Office

Figure 5.3 Basic wind speed, V

the responsiveness of the structure under consideration and this is a function of structure mass, stiffness, damping and geometry. The British Standards document CP3: Chapter V: Part 2 'Wind loading' uses a set of factors reflecting the topography, ground roughness and design life of a structure to condition a basic fifty year return wind speed to that with a small probability of being exceeded in a particular location in the design life of a structure. In other countries, such as Canada, dynamic response characteristics are incorporated in the wind action assessments on a full probability basis. This trend is under review in the UK and is likely to be reflected in future issues of British Standards for wind loading.[4] For simple unresponsive structures of small scale, reasonable wind loading values for design purposes can be obtained by direct application of CP3: Chapter V 'Wind loads' methods. For large or complex structures direct overall application of this procedure may result in conservative overall wind design forces for uplift and overturning whilst possibly yielding non-conservative local values. To provide an accurate wind load assessment in the case of a large or complex structure it is therefore recommended tests be carried out to avoid uncertainties.[7,8]

(*c*) *Earthquake* – In the UK there is a small earthquake risk in most locations (fig. 5.4) but this is not the case in many countries. Even where the risk of a severe earthquake (table 5.2) is fairly high for a return period of fifty years or so a designer may have to balance the additional costs of non-collapse design of falsework for such an event against the probability of occurrence and the insurance costs.[9] If it is decided to incorporate earthquake action as one of the design criteria for a falsework system then this will generally be carried out in terms of non-collapse rather than that of serviceability after the event.

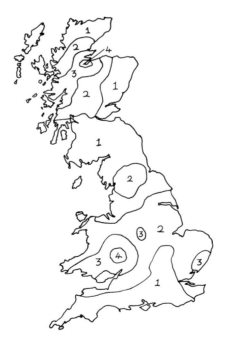

Figure 5.4 Seismic zones for Great Britain

In general terms seismic forces are predominant at low frequency in horizontal directions but the higher frequency vertical vibration forces cannot be ignored since they tend to hammer vertical supports, possibly resulting in buckling of columns or dislodgement of fixtures, and this vertical vibration usually causes a considerable reduction in the frictional resistance at horizontal interfaces which are frequently encountered in many falsework systems. This reduction in horizontal shear resistance in conjunction with the major horizontal seismic forces can impose extremely severe conditions in a falsework system even in a minor earthquake. It cannot be attempted here to illustrate the various methods of earthquake analysis used to evaluate seismic forces but for initial design assessments static equivalent horizontal shear forces on a system can be found from an equation of the form:

$$V = RWC$$

where R = seismic risk coefficient (e.g. 0.2 for 20% 'g' ground acceleration probability)

W = structure weight including superimposed loading

C = $0.05/\sqrt[3]{T}$

T = fundamental horizontal period of vibration (say $0.2H/\sqrt{D}$ as a first estimate for bridge falsework of height 'H' and least lateral dimension 'D'.

(d) *Flood* – Where falsework structures are partially immersed or founded on river beds or coastal waters there is a probability[10] of flooding or severe wave action. The magnitude of flooding or wave action likely to occur during the life of a falsework structure can often be determined from past records. Detailed action may then comprise a combination of vortex action, buoyancy, current forces, scouring and possibly impact from debris and upward slam forces on surfaces above the mean water level. These forces are a function of structure geometry, current velocity, wave heights and periods, thus requiring detail study for specific structures. Depending upon the probability of occurrence, secondary protection or streamlining of falsework systems can be of great advantage in restriction of damage and in minimising loss of construction time.

(e) *Snow* – Some general guidance is given in CP3: Chapter V and BS 5975 regarding snow loadings. The importance of this loading depends upon the type of system under investigation. In the case of a heavy falsework structure designed to support the construction of a bridge deck then the general 1.5 kN/m² allowance for access and working areas is generally sufficient to allow for any snow loading during construction. In the case of lightweight structures where even moderate snow loading over a large area could represent a major force action then careful consideration of snow action is required if an economic solution is to be found. The factors required to be considered in an assessment of probable snow loading include the structure design life, snow records for the site under consideration, structure geometry and wind direction for possible retention of snow, drifting or shedding, the time of year construction takes place if the design life of the structure is less than one year and any systems employed for reduction or removal of snow loading.

(f) *Ice* – In some sites, normally of an exposed and elevated type, icing of a structure during periods of severe winter weather can occur. Where this is a probability consideration of ice of up to 25 mm in thickness on structural members may be necessary and the possibility of complete icing over of openings in lattices may have to be analysed. In some instances the wind forces derived from CP3: Chapter V: Part 2, Appendix F are slightly less when icing is assumed than when considered absent since the design wind speed is reduced with icing and although the effective area of a structure is greater with icing, significant reductions in design wind speeds during icing at elevated and exposed northern sites could be questioned.

Table 5.2 Earthquake intensity and magnitude scales and UK seismic zones

Modified Mercalli scale (abridged 1931)	Rossi-Forel equivalent	Richter magnitude	Assumed approximate acceleration (m/s²)	UK seismic zone	General effect
I	I	← 2 →			Felt only by a few in especially favourable circumstances.
II	I–II				Felt only by a few at rest, especially on upper floors. Delicately suspended objects may swing.
III	III	3		1	Felt quite noticeably indoors, especially on upper floors. Many do not recognise effects as earthquakes. Standing cars may rock slightly. Vibrations seem like a lorry passing.
IV	IV–V	← 4 →	0.01–0.0255		Daytime felt indoors by many, outdoors by few. At night some awakened. Dishes, windows, doors disturbed. Walls creak. Standing cars rock considerably. Sensation like a heavy lorry striking a building.
V	V–VI		0.0255–0.051	2	Felt by nearly all. Many awakened. Some dishes, windows, etc. broken. Some cracked plaster. Unstable objects overturned. Trees, poles disturbed. Pendulum clocks may stop.
VI	VI–VII		0.051–0.102	3	Felt by all. Many frightened and run outside. Some movement of heavy furniture. Few cases fallen plaster and slight damage to chimneys.
VII	VIII	6 →	0.102–0.255		All run outdoors. Negligible damage in well-designed and constructed buildings, slight–

Intensity	Magnitude	Acceleration	Description
VIII	—	—	moderate in normal buildings, considerable in badly built structures, some broken chimneys. Shock noticed by drivers in vehicles.
VIII–IX	7	0.255–0.51	Slight damage to aseismically designed structures; considerable in ordinary and partial collapse; great damage in poorly built. Panel walls thrown from framed structures. Fall of chimneys, columns, walls. Heavy furniture overturned. Vehicle drivers disturbed. Small sand and mud boils.
IX		0.51–1.02	Considerable damage to specially designed structures. Well-designed frame structures out of alignment. Partial collapse of substantial buildings. Buildings shifted from foundations. Ground conspicuously cracked. Underground pipes broken.
IX+			
X		1.02–2.55	Some well-built wooden structures destroyed. Most masonry and frame structures destroyed with foundations. Rails bent. Landslides considerable from river banks. Sand and mud shifted. Water slopped over banks.
X	8	2.55–5.10	Few, if any, structures left standing. Bridges down. Broad fissures in ground. Underground pipe lines completely out of service. Rails greatly bent. Earth slumps and landslips on soft ground.
XI	8+	>5.10	Damage total. Waves seen on ground surface. Objects thrown upwards into the air. River beds displaced. Numerous and extensive landslides.
XII			
X+			

Notes to table 5.2
1. The acceleration figures are an indication rather than an accurate measurement.
2. It will be seen that the acceleration roughly doubles for each class in the Modified Mercalli scale.
3. The Richter magnitude column values are independent of surface structure existence, type or damage.
4. The UK seismic zone comparisons are for a 200 year return.

5.2.3 Constraint effects

(a) *Creep and shrinkage* – The effects of creep and shrinkage are usually long term and dealt with by the insertion of movement gaps but in many instances precamber of formwork and supporting falsework structures is carried out to minimise possible deflections from this source such that visually the structure remains unimpaired and possible reduced serviceability and possible water pooling are eliminated.

(b) *Temperature* – Differentials of 25°C or more between sunny and shaded sides of structures are possible, leading to relative movements of about 40 mm for a 150 m length and an expansion coefficient of $11 \times 10^{-6}/°C$. Where this could be a problem the scheme and timing of concrete placement may have to be designed specifically to minimise adverse effects.

(c) *Settlement* – Even small rotations or differential movements can radically alter load distributions in falsework structures, can cause misalignment of formwork leading to incorrect finished structural geometry and can significantly influence the ultimate capacity of the falsework structure.

(d) *Out of plumb* – If erection of a falsework structure is not closely controlled then small angular inconsistencies of support columns or uneven footings can result in considerable out of plumb of a structure. This is not readily quantified at the design stage but can alter structural stability, cause considerable bending stress in verticals and redistribution of forces and can greatly reduce the capacity of a falsework structure. If problems of this nature are anticipated then the effects can often be minimised by using simply supported systems rather than continuous beams in order to accommodate support movements.

5.3 ALTERNATIVE FALSEWORK SCHEMES FOR CONSTRUCTION OF A BRIDGE

In the previous sections an attempt has been made to describe the main reasons for the choice of elastic design and analysis for most falsework structures and to give a global view of the forms and major influences of force actions generally considered in the design of falsework structures. In line with this, elastic analysis and design is employed in dealing with force actions on the falsework systems to be considered in this section.

The general arrangement of a highway bridge to be constructed across a valley, which contains a river or a minor roadway, is shown in fig. 5.5. To illustrate the development of appropriate design forces and methods which could be employed to provide support structures and erection procedures for such a bridge a number of false-work design procedures are considered below.

Common to each scheme is the construction of the bridge piers and abutments which are assumed to be completed prior to commencement of placement of the deck structure. Two obvious systems, namely shuttered lifts and alternatively, continuous pour with sliding shutters, could be used to construct the piers. Where piers are uniform or evenly tapered in cross section, are of significant height and do not contain complex openings or changes in reinforcement, then moving shutters and continuously

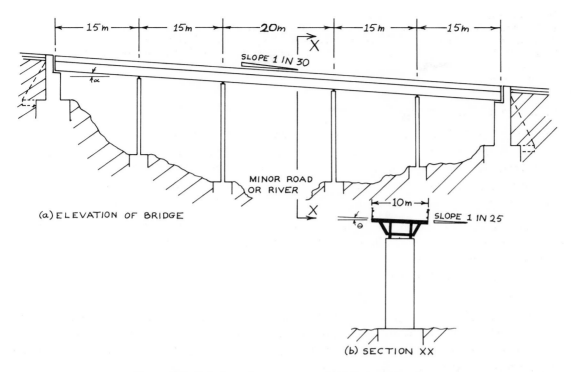

Figure 5.5 Reinforced concrete road bridge over a valley

poured concrete presents an economical construction method (fig. 5.6). If however
the piers and abutments are relatively short, contain complex reinforcement, change
shape or contain openings then often shuttered lifts can be more economical than
continuous pour methods.

5.3.1 General bridge falsework scheme 1 – Centre span

The falsework employed in this scheme includes the use of fabricated trusses and
towers, Universal sections and scaffold poles to form the support structure for casting
the bridge deck. For details of the formwork design reference should be made to
chapter 2.

Construction and environmental loads

Many loadings may have to be considered in the design or assessment of a falsework
structure as indicated in section 5.2 but here for simple illustrative purposes only
self-weight, imposed loads and environmental loads are considered.

Detail consideration of the weights of the centre span deck, ancillary works and
equipment, formwork, deck support falsework structures, personnel involved, stacked
materials and temporary heaping of concrete[1,11,13] yields a total vertical loading from
the centre span of 5230 kN. (Note – 1 kg acting vertically is equivalent to 9.81 N.)

The wind forces to be resisted by the falsework structure are estimated here for a
region where there is a very low probability of icing of the falsework structure during
its design life. The deck of the bridge is taken as 14 m above the valley floor and a
fifty year wind gust velocity of $V = 50$ m/s is to be adopted.[1,3] The valley is not
particularly sheltered and neither does it funnel the wind and therefore a topography
factor equal to unity is appropriate ($S_1 = 1.0$). The site contains scattered wind breaks
such that a surface factor of 0.98 taken from fig. 4 of reference 1 is appropriate
($S_2 = 0.98$). This factor is deemed to modify the basic wind speed for ground roughness

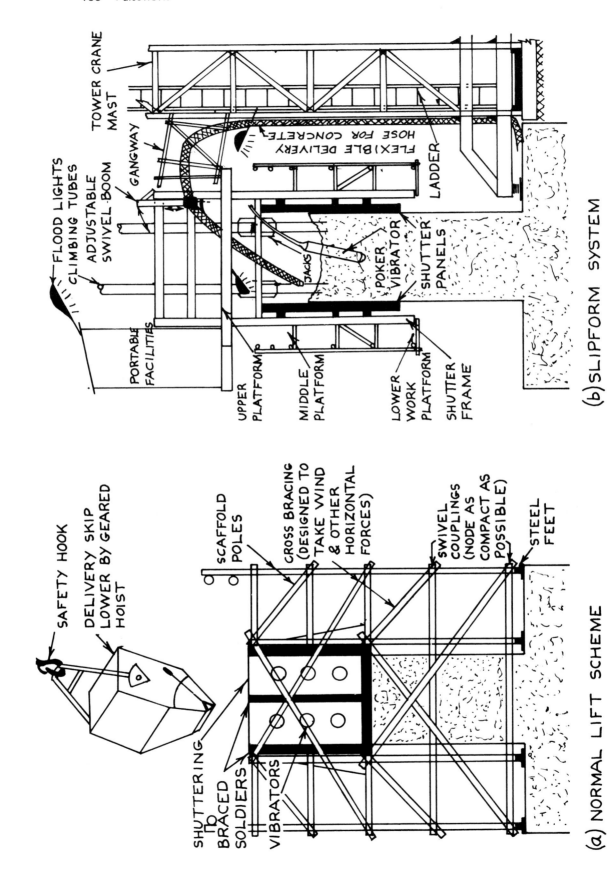

SAFETY HOOK

DELIVERY SKIP LOWER BY GEARED HOIST

SCAFFOLD POLES

CROSS BRACING (DESIGNED TO TAKE WIND & OTHER HORIZONTAL FORCES)

SWIVEL COUPLINGS (NODE AS COMPACT AS POSSIBLE)

STEEL FEET

SHUTTERING TO BRACED SOLDIERS

VIBRATORS

(a) NORMAL LIFT SCHEME

FLOOD LIGHTS

CLIMBING TUBES

ADJUSTABLE SWIVEL-BOOM

GANGWAY

TOWER CRANE MAST

FLEXIBLE DELIVERY HOSE FOR CONCRETE

LADDER

JACKS

POKER VIBRATOR

SHUTTER PANELS

PORTABLE FACILITIES

UPPER PLATFORM

MIDDLE PLATFORM

LOWER WORK PLATFORM

SHUTTER FRAME

(b) SLIPFORM SYSTEM

Figure 5.6 Falsework for construction of the bridge piers

and the structure size and height (see also appendixes A and D of reference 3, especially if a structure is adjacent to cliffs or embankments). The basic wind speed is further modified depending upon the design life of the structure and here the maximum feasible design life, allowing for possible down time from inclement weather or other delays, is taken as twenty months. This results in a statistical design life factor for wind of 0.77 ($S_3 = 0.77$).

Employing the above values the design wind speed is given by:

$$V_s = VS_1 S_2 S_3 = 38 \, \text{m/s}$$

From table 12 of reference 1 or table 4, reference 3 the dynamic wind pressure for a design wind speed of 38 m/s is:

$$q = 885 \, \text{N/m}^2$$

Note – For rapid design purposes a temporary 'q' value of 1 kN/m² may often be adopted and the results scaled at a later date when an accurate 'q' value is known.

Once an estimate of the effective frontal area of falsework or component (A_e in m²), the force coefficient (C_f) related to the shape of the member or frame subject to wind action and the shielding factor (ζ) are known then the maximum probable wind forces likely to act during the life of a falsework structure can be obtained from:

$$W_m = qA_e C_f \zeta$$

For the structure under consideration the force coefficients for the various structural components are taken as:

deck $= 1.0$
personnel balustrade $= 1.20$
windward falsework lattice girder $= 1.88$ (with a shielding factor of 0.41 for the other girders).

Summation of the horizontal wind force acting at right angles to the bridge deck is therefore:

$$\Sigma qA_e C_f \zeta = 196 \, \text{kN}$$

The deck of the bridge is sloping in two directions (fig. 5.5) therefore basic design loadings on the centre span falsework, including self-weight of all temporary works, can be summarised as:

Force acting vertically downwards (no wind uplift) $= 5230 \, \text{kN}$
Horizontal force acting along the bridge ($W \sin \alpha$) $= 144 \, \text{kN}$
Horizontal force acting across the bridge ($W \sin \theta$) $= 209 \, \text{kN}$
Wind force acting across the bridge $= 196 \, \text{kN}$
Allowance for wind along the bridge and piers and
 acting at deck level $= 71 \, \text{kN}$

Snow loading on the main deck formwork is already accounted for within a general vertical loading allowance of 1.5 kN/m² for general working and access (BS 5975 clause 28.3). To account for an improbable possibility of drifting of snow to 1 m deep on the personnel walkways an extra 0.8 kN/m² of vertical loading is assumed at all such locations for added safety. This adds 32 kN to the vertical loading and approximately 1 kN to the horizontal force in each direction.

If the structure were to be deployed in a climatic region where winter conditions may cause icing on the structure then ice up to 25 mm thick on tubes and lattice members would be considered and in some instances complete filling over of lattice openings could occur. In such circumstances alteration would be made to the wind loading and the mass of the ice would require to be accounted for. In some regions in Norway, cycles of snow fall followed by partial melting, refreezing and further snow falls can lead to massive build-ups of ice and snow which require careful consideration. For the centre span under consideration additional loadings from icing would amount to:

Vertical loading from ice   = 303 kN
Horizontal component along centre span   =   10 kN
Horizontal component across centre span   =   12 kN

The wind force on the falsework would also have to be modified according to appendix F of reference 3 for a probable reduced design pressure but an increased effective area. In the UK it is very unlikely that icing would occur during vital casting stages of construction but again in some northern countries there is a higher probability of significant icing and high wind occurring simultaneously.

Since icing is improbable in the region under consideration such loading will be ignored in this design example and the overall design loadings become:

Total vertical force from the centre span to the falsework support towers   = 5262 kN
Total force across the bridge centre span deck and falsework   =   406 kN
Total force along the bridge at the centre span deck level   =   216 kN

Centre span deck support falsework

In scheme 1 (figs. 5.7 and 5.8) the formwork and deck are supported on eleven Warren girders, details of which are given in figs. 5.9 and 5.10. The uniformly distributed load on the most heavily loaded girder is 30 kN/m.

Before considering the analysis and detail design checking of the lattice girders it is

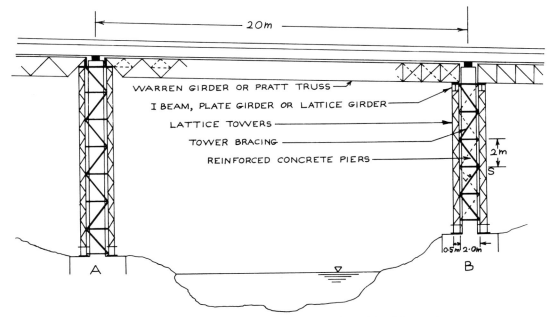

Figure 5.7 Elevation of general scheme 1 for bridge falsework

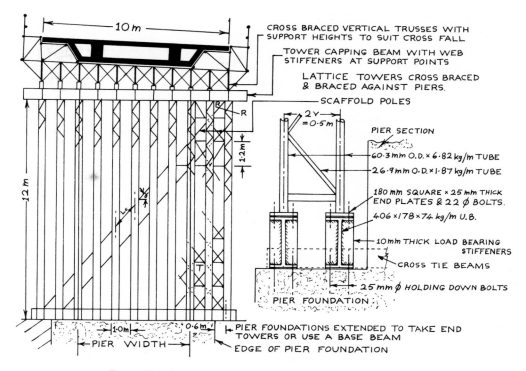

Figure 5.8 Section through general falsework scheme 1

worth noting the following general points regarding the design and deployment of such girders:

(a) The girders must be placed vertically to avoid inducement of lateral bending forces.

(b) The end support feet of the girders must be angled or packed to take up the cross fall of the bridge deck.

(c) The load carrying capacity, deflections and element and overall stability of the girders require to be checked. Precamber of the girders may be used to allow for deflections under load.

(d) The lateral stability of the parallel girders to the horizontal design forces has to be ensured by means of bracing or moment resisting connections. For falsework structures bracing would generally be preferred for this purpose.

(e) The neutral axes of all elements which meet at a node should cross at a single point. If this is not the case then extra shear and bending of the elements is induced and additional stresses have to be carried by the connections. For scaffold structures this is reflected in reduced allowable stress values for the scaffold tubes. Allowable stresses in welds, based on throat thicknesses, are given in BS 449, clause 53a. For bolted joints checks have to be made for shear in the bolts, gussets and members and bearing and tension in the gussets and members.

(f) For falsework structures mild steel grade 43A is generally specified and should only be exceeded where a higher grade steel is specified. If a check on steel members of unspecified grade has to be carried out then grade 43A should be assumed.

(g) Where a load test is to be used as a means of checking a girder then the factor of safety against collapse in the most unfavourable simulated on-site load conditions is required to be two or greater.

Simple analysis and design check of the lattice girders

Lattice frames of the size under consideration are generally of welded construction or have multi-bolted joints. For an 'exact' analysis they would not be considered to be pin-jointed structures since a certain degree of joint fixity exists but in most instances the bracing elements are sufficiently slender to render an analysis on the basis of a pin-jointed structure accurate to within a few percent of the 'exact values'.

In the analysis of lattice girders there are a number of general rules which are worthy of note. The top and bottom booms absorb the overall bending moment on the girders while the bracing absorbs the shear forces. Maximum bending moments and deflections generally occur near mid-span for simply supported girders and maximum shear forces occur at the supports. For the uniformly loaded girders in the example under consideration identification of the most highly stressed elements can be carried out as follows:

$$\text{Support reaction} = wL/2 = R = 270\,\text{kN}$$

(For other load patterns take moments about one end support of the girder.)

$$\text{Force in element AB (fig. 5.9)} = \frac{w(L/2 - z)}{\cos \phi} = 277\,\text{kN (tension)}$$

To find the force in element DE take a section XX through the lattice frame and consider the moments acting about point C for the left-hand part of the girder from which the force in element DE

$$= \frac{x(R - wx/2)}{h} = 693\,\text{kN (tension)}$$

where h is the effective depth of the lattice girder. To find the force in element FC take a section YY through the frame and moments about point D for the left-hand part of the frame. From this the force in element FC

$$= \frac{y(R - wy/2)}{h} = 702\,\text{kN (compression)}$$

The force in BG can be found by taking a section ZZ and summing the vertical force components for the left-hand part of the frame. The force in element BG is then

$$= \frac{w(L/2 - z)}{\cos \beta} = 277\,\text{kN (compression)}$$

If we had wished to analyse all the parallel girders and assess their interaction simultaneously then a three-dimensional analysis could have been employed but only rarely can such a degree of sophistication be justified for simple falsework systems.

In addition to the axial forces, those elements directly supporting formwork are also subject to bending forces. If the top boom is continuous and an intermediate element, say FC, is considered then the mid-span moment on FC $= wl^2/24 = 5\,\text{kNm}$ and at F or C $= wl^2/12 = 10\,\text{kNm}$.

The effective lengths of the girder elements for assessment of their load carrying capacity and stability are obtained from BS 449 or BS 5975 which also contains guidance on the fabrication, alignment and minor damage tolerances. For the parallel girders cross braced as shown in figs. 5.7 to 5.9 the effective lengths of the elements are of order 0.70 × actual length for interior braced boom elements and 0.85 × actual length for other elements. For simplicity and to avoid reduction in safety the effective length is taken as $L_e = 0.85$ × actual length as the general case for the braced lattice girder elements.

The allowable axial compression in a strut based on the maximum value of L_e/r (the effective length of a member for a particular axis of bending divided by the radius of gyration of the member cross section for the same axis of bending) is given in table 17 of BS 449 and table 22 of BS 5975 for structural sections other than scaffold tubes connected by couplers for which reduced allowable values are given in table 23 of BS 5975.

Note – The effective length of an element may not be the same about all axes of bending and therefore several values of L_e/r may require to be calculated for some elements in order to obtain the correct value for permissible axial stress or load.

The allowable bending induced compression in the flange of a beam element is given in table 3 of BS 449 or table 21 of BS 5975. The assessment is made on the basis of the ratios L_e/r_y and D/T (the ratio of beam depth to flange thickness).

The allowable axial tension in an element is goverened by the material properties, the net cross-sectional area of the element and the deflection criteria.

For members subject to both axial compression and bending the quantity $f_c/p_c + f_{bc}/p_{bc}$ must be less than unity

where
f_c = average calculated compressive stress
f_{bc} = resultant compressive stress from bending about the vertical and horizontal axes
p_c = allowable compressive stress
p_{bc} = allowable bending compressive stress.

For members subject to both axial tension and bending the quantity $f_t/p_t + f_{bt}/p_{bt}$ must be less than unity

where
f_t = calculated axial tensile stress
f_{bt} = resultant tensile stress from bending about vertical and horizontal axes
p_t = permissible axial tensile stress from table 19, BS 449
p_{bt} = allowable tensile stress in bending from table 2, BS 449.

In general the philosophy of the design should be to design the overall structure for elastic action but to detail the structure to take plastic action.

An assessment of the individual elements of the Warren girders can now be made starting with the top boom. The most highly stressed top boom element is FC which is compressed by a force of 702 kN from the vertical loading and a force of 26.6 kN from horizontal forces acting on a pair of braced girders. The bending moment in FC from the vertical uniformly distributed load is 10 kNm and from the horizontal forces is 2.173 kNm. If a 203 × 203 × 52 kg/m Universal Column top boom is used then the total compressive stress in FC is

$$f_c = \frac{\text{load}}{\text{cross-sectional area}} = \frac{728.6 \times 10^3}{66.4 \times 10^2} = 109.7 \text{ N/mm}^2$$

and the bending compressive stress is

$$f_{bc} = \frac{M_v y}{I_{xx}} + \frac{M_h x}{I_{yy}} = \frac{M_v}{Z_{xx}} + \frac{M_h}{Z_{yy}} = \frac{10 \times 10^6 \times 103}{5263 \times 10^4} + \frac{2.173 \times 10^6 \times 102}{1.77 \times 10^7} = 32.1 \text{ N/mm}^2$$

To obtain allowable values evaluate $L_e/r_y = 0.85 L_{FC}/r_y = 32.17$; $D/T = 16.5$ therefore $p_c = 142 \text{ N/mm}^2$; $p_{bc} = 165 \text{ N/mm}^2$ and the quantity $f_c/p_c + f_{bc}/p_{bc} = 0.967$. This is less than but close to unity and therefore the top boom is of an economic and suitable size for the maximum design loadings.

For shear transfer at A to the support through the web of the beam the shear stress in the web is

$$f_q = \frac{\text{vertical shear force}}{\text{web area}} = 167 \, \text{N/mm}^2$$

For grade 43 'I' sections the maximum allowable web shear stress from table 10 of BS 449 is 115 N/mm² but from BS 5975: Appendix A is 100 N/mm². For tubes the shear area is taken as half the cross-sectional area and for rolled 'I' beams is taken as the beam depth times the web thickness. Where only flexural and shear forces are involved BS 5975: Appendix A also allows 115 N/mm² shear stress in 'I' beam webs.

It is apparent from the above that the shear stress would be greater than the allowable value. The design solution is to detail the diagonal to be fixed to a gusset welded to both the support leg and the boom plus the provision of web stiffeners in the top boom.

The most highly stressed bottom boom element is DE in which the total axial tensile force from vertical and horizontal actions is 719 kN. If the bottom boom is a 219.1 mm outside diameter tube × 41.6 kg/m with no holes bored then the axial tensile stress is

$$f_t = \frac{\text{tensile force}}{\text{net cross-sectional area}} = \frac{P}{A} = 135.4 \, \text{N/mm}^2$$

and from table 19, BS 449 or Appendix A, BS 5975 the allowable axial tension on the net sectional area is $p_t = 155 \, \text{N/mm}^2$.

The axial tensile force in diagonal element AB is 277 kN and if an unperforated 114.7 mm outside diameter tube × 16.8 kg/m is used then the tensile stress is $f_t = 129.4 \, \text{N/mm}^2$ and again the allowable tensile stress is $p_t = 155 \, \text{N/mm}^2$. The axial compressive force in diagonal BG is 277 kN and if a 114.7 mm outside diameter tube × 16.8 kg/m is used then the compressive stress is $f_c = 129.4 \, \text{N/mm}^2$. For element BG the ratio

$$\frac{L_e}{r_{min}} = \frac{0.85 L_{BG}}{38.2} = 44.5$$

therefore the allowable compressive stress is $p_c = 136.5 \, \text{N/mm}^2$, from table 17, BS 449 or table 22, BS 5975.

Sometimes it is an advantage to further triangulate a Warren girder as shown in fig. 5.7 when the top boom is subject to bending and the diagonals weak axis is simultaneously braced. Additionally if two top booms and a single bottom boom are used the lateral stability can be increased but in our scheme cross bracing between the parallel girders is incorporated.

We will now consider the form and extent of cross bracing required between the Warren girders. The maximum design force at right angles to the girders in this case is that shown in fig. 5.9 which also illustrates a suitable cross-brace system. To assess the bracing capacity let us consider that the two end girders braced together have to take all the wind loading on the girders plus one fifth of the total lateral load transferred to the girders from the deck and ancillary works. It is assumed that lateral loading may act from either side of the bridge to ensure safety, simplicity and to achieve symmetry of falsework to avoid erection errors. The Warren girders seen in plan may be analysed in the same manner as the main vertical girders by means of the method of sections or other simple analysis. The resulting maximum forces in the bracing are then found as follows. Ignoring any node compression transfer relaxation, including lateral restraint of the main girders, the maximum feasible tensile force in element A1 is 61 kN

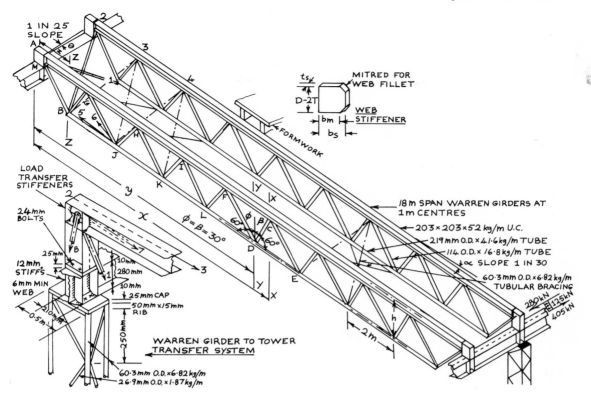

Figure 5.9 Details of deck support structure for general scheme 1

and compressive force in element 1G is 61 kN. The additional shear at 'A', taken through the top boom flanges and stiffeners, is 43.3 kN. If unperforated 60.3 mm outside diameter $\times$ 6.82 kg/m tubes are used for the bracing then for element 1G the compressive stress is $f_c = 70.1$ N/mm². For 1G the maximum ratio

$$\frac{L_e}{r_{min}} = \frac{0.85 L_{1G}}{19.6} = 86.7$$

(and possibly a factor of 0.70 could be used instead of 0.85), therefore the allowable compressive stress is $p_c = 95$ N/mm².

A girder is provided to transfer the lattice beam forces to the support towers. The maximum vertical force transferred from a deck support girder is 270 kN and the maximum horizontal force transferred is 43 kN. Maximum moment transfer is calculated to be approximately 20 kNm. In the vertical legs the bending moment acts about the weak axis therefore the maximum feasible bending stress is

$$f_{bc} = \frac{M_x}{I_{yy}} = \frac{20 \times 10^6 \times 102}{1770 \times 10^4} = 115.3 \text{ N/mm}^2$$

for the worst feasible load distribution. The maximum compressive stress is $f_c = 40.7$ N/mm² and horizontal shear stress is 6.5 N/mm². The values of L_e/r and D/T result in maximum allowable bending and axial stresses of $p_c = 155$ N/mm² and $p_{bc} = 165$ N/mm². Allowable shear stress is $p_q = 100$ N/mm². Using these stress values

$$f_c/p_c + f_{bc}/p_{bc} = 0.961$$

which is less than unity as required. The allowable equivalent stress is sufficiently high at 230 N/mm² not to be even approached by the stress combinations of clause 14c of BS 449 for the design stresses calculated above.

To check for possible web buckling in the capping beam the maximum load transfer through the unstiffened web is found from:

$$p = p_c t B$$

where p_c = allowable axial stress for a strut of slenderness ratio = $2.4D/t$, assuming the web ends fully restrained (see also Appendix K, BS 5975)
t = web thickness
B = $D/2 + t_p + l_b$ for end bearings of simply supported beams
= $D + 2t_p + l_b$ for intermediate bearings or loads
D = beam depth
t_p = bearing and/or flange plate thickness
l_b = length of stiff portion of bearing (not greater than $D/2$ for simply supported beams and D for continuous beams)

Therefore, for no web stiffener, the maximum load transfer to the tower capping beam for end Warren girder support is a minimum of

$$61 \times 6 \times (300/2 + 12.5 + 203) \times 10^{-3} = 134\,kN$$

To check for possible web crushing in the capping beam the maximum load transfer for simply supported bearings and no web stiffeners is given by:

$$p = 190t \sqrt{3} \,[(D-d)/2 + t_p]$$
$$d = \text{web depth between root fillets}$$

and for continuous beam bearings:

$$P = 190t \times 3.46 \,[(D-d)/2 + t_p]$$

Therefore, for no web stiffeners and for adequate safety against the occurrence of web crushing maximum load transfer to the tower capping beam

$$= 190 \times 6 \times \sqrt{3} \times (40/2 + 12.5) \times 10^{-3} = 65\,kN\,(minimum)$$

The design of the load-bearing stiffeners for the capping beam (fig. 5.9) is now carried out by first calculating the effective area for resisting compression which is equal to the cross-sectional area of the stiffeners plus a $20t$ length of the web to each side of the stiffeners

$$= 20t^2 \times 2 + 2b_s t_s = 3865\,mm^2$$

where b_s and t_s are the cross-sectional dimensions of a stiffener. The compressive stress in a stiffener

$$= \frac{\text{load}}{\text{effective area}} = 95\,N/mm^2$$

The second moment of area of the stiffeners is

$$I_s = \frac{t_s(2b_s)^3}{12} = 8.0 \times 10^6\,mm^4$$

and therefore the radius of gyration is

$$r_y = \sqrt{(I_s/A_e)} = 45.5\,mm$$

The effective length of the stiffeners is approximately $L_e = 0.7d$, therefore

$$\frac{L_e}{r_y} = \frac{0.7d}{45.5} = 4.0$$

From table 17, BS 449 or table 22, BS 5975 the allowable compressive stress is, $p_c = 155\,\text{N/mm}^2$.

$$\text{Bearing stress} = \frac{\text{vertical load}}{2b_m t_s} = \frac{270 \times 10^3}{2 \times 85 \times 12} = 132\,\text{N/mm}^2$$

where b_m is as shown in fig. 5.9. From table 9, clause 22 of BS 449 or Appendix A of BS 5955, $f_b = 190\,\text{N/mm}^2$.

If a tower is not positioned beneath each girder support point then bending of the capping beam has to be considered. For plate girders with stiffened webs the stiffeners have to be spaced according to clause 28b(i), BS 449 and the permissible shear stress is given by table 12, BS 449 using the ratio d/t. If bending stress f_{bc} in a beam is greater than $0.6p_{bc}$ at a load point then $f_r/p_b + f_{bc}/p_{bc} + f_c/p_c$ must be less than 1.6 where f_r = vertical load or reaction at the point/Bt.

The feet of the Warren girders have to be set such that the girders remain vertical. The girder support legs to capping beam connections have to be capable of transmitting the shear forces and the bending moments (fig. 5.9).

Splices of the Warren girders should be planned such that they occur away from positions of maximum stress.

Temperature changes cause expansion and contraction of the deck support system and provision can be made for this at the feet of the Warren girders using slotted bolt holes and hard plastic or rubber spacers.

Tower design assessment

The most heavily loaded tower supports a load of 280 kN at its base divided equally between each of its legs by means of the capping and footing devices (fig. 5.9). The total loading on a tower results from a combination of the vertical load and forces induced in the towers in resisting the lateral forces across and along the bridge.

Stability of the towers to the horizontal force along the deck is obtained by a combination of bracing against the piers and bracing between the towers outside the line of the piers. The maximum longitudinal design force at one pier results from half the centre span force and half of the force generated by the adjacent 15 m span. The result is a design force at the top of a pier of $216 \times 1.75/2 = 189\,\text{kN}$ during worst probable construction conditions. If the falsework towers and inter-tower longitudinal bracing are required to take all this force then, considering only the towers outside the line of the piers as the booms of Warren girders resisting the longitudinal forces, the maximum vertical force in a tower from this source is

$$\frac{189 \times H}{6 \times L_a} = \frac{189 \times 12}{6 \times 2.5} = 151\,\text{kN}$$

This figure is arrived at by assuming only six pairs of towers attracting this loading and ignoring any shear lag across the rows of towers. Analysis, say by the method of sections, as for the deck support girders of two towers diagonally braced together provides the forces in the bracing elements. The maximum compressive force is 63 kN in a diagonal if the pattern of bracing shown at pier 'A' (fig. 5.7) is adopted. Normally in a falsework structure, to avoid the use of heavy bracing and to provide an added measure of safety, cross bracing as shown at pier 'B' (fig. 5.7) is used and after the contribution of the compressive diagonal is ignored. This results in 63 kN tension in a diagonal and 31.5 kN compression in a horizontal element. For unperforated 60.3 mm outside diameter × 6.82 kg/m tubes used for inter-tower brace elements the maximum tensile stress is $f_t = 72.5$ N/mm² and compressive stress is $f_c = 36.3$ N/mm². Welded end lugs with plain barrel piers or bolts can be used to transfer the full tensile capacity of a tube with no reductions (fig. 5.9). For the compressive elements the ratio of effective length to radius of gyration is,

$$\frac{L_e}{r} \simeq \frac{2200 \times 0.85}{19.6} = 95.5$$

which results in an allowable compressive stress of $p_c = 85$ N/mm².

Stability of the towers to the horizontal forces across the bridge may be obtained by numerous forms of bracing (figs. 5.8 and 5.10). In this scheme triangulation by means

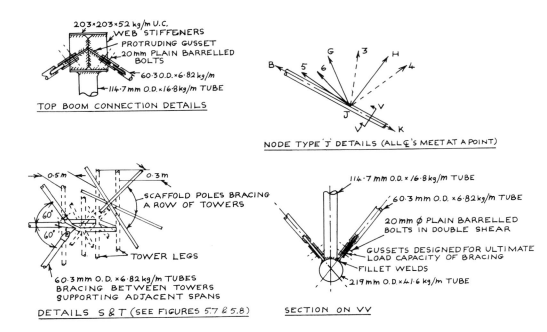

Figure 5.10 Bracing details for scheme 1

of scaffold poles is adopted to resist the horizontal forces on a row of towers and increase their stability. The pier inertia about its stiff axis could also be used if sufficient curing has been allowed to take place. The horizontal force at the top of a row of towers including an allowance for wind loading on the towers = 406/2 = 203 kN. For equal load transference to all towers the maximum compressive force in a tower in the braced system from this source is approximated by taking a section RR (fig. 5.8) which yields the force = 203/10 tan δ ≃ 203/10 tan 40° = 24.2 kN. The force in a bracing element of scaffolding is a feasible maximum of 20.3/cos 40° = 26.5 kN. Transfer of a force of this magnitude would require multi-couplers or welding. To allow for possible ineffective members such that a diagonal element were forced to take compression then the compressive stress would be:

$$f_c = \frac{26.5 \times 10^3}{557} = 47.6 \text{ N/mm}^2$$

The maximum value of L_e/r is given by:

$$\frac{L_e}{r} = \frac{L_c}{r} = \frac{1560 \text{ max}}{15.7} = 99.4$$

Therefore for a minimum factor of safety of 2.0 the allowable compressive stress from table 23, BS 5975 is, $p \simeq 70 \text{ N/mm}^2$.

From the vertical load and the longitudinal and transverse forces on the bridge falsework the total compressive stress in a tower leg is a maximum feasible of:

$$f_c = \frac{\text{tower vertical force}}{(\Sigma A \text{ of tower vertical tubes})} = \frac{458 \times 10^3}{4 \times 8.69 \times 10^2} = 131.76 \text{ N/mm}^2$$

for 4×60.3 mm outside diameter $\times 6.82$ kg/m tubes as the vertical elements in a tower. The effective length of these elements between the tower internal bracing, $L_e = 0.7L = 350$ mm (figs. 5.7 to 5.9), therefore $L_e/r = 17.9$ and for this value the allowable compressive stress is $p_c = 148 \text{ N/mm}^2$. This assumes that compressive buckling of a tower would take the form of buckling of the most highly stressed strut but for the towers inside the line of the piers, buckling of a complete tower away from a pier between the capping and base beams is possible. The effective length of a complete tower is approximately $1.0H$ if the capping and base beams are properly secured. The towers are symmetric about each axis to avoid the prospect of incorrect placement. The minimum practical value of second moment of area of a tower about a central axis parallel to one face of a tower is approximated by, $I = 4 \times A_{leg} \times v^2/4$ and the radius of gyration is approximated by, $\sqrt{(I/4A_{leg})}$ or directly by $r_x = r_y = v = 250$ mm. Therefore $L_e/r = 12000/250 = 48.0$ which results in an allowable compressive stress of, $p_c = 135 \text{ N/mm}^2$. A more exact value for the critical buckling load for a tower about either axis can be found from a matrix stability analysis (reference 14).

In the above calculations, out of plumb of a tower has not been considered. Even a small deviation from vertical positioning of a tower would alter the load distribution and can greatly reduce the safe load carrying capacity.

Tower base system

In each row of towers only the end towers fall outside the pier foundations. One method of solution is to extend the pier foundations and bolt the towers individually to the foundation to take the shear and overturning moments transmitted to the foundations. Another method is to bolt the towers to beams anchored to the pier foundations and cantilevered out at each side of the pier bases. For these beams the design of web stiffeners to resist web crushing and buckling is similar to that for the capping beam.

To check the bending stresses in the cantilevered section the maximum bending moment is obtained from:

$$M_{max} = \text{force} \times \text{cantilever length} = \frac{458 \times 0.6}{2} = 137.5 \text{ kNm}$$

If twin $406 \times 178 \times 74$ kg/m Universal Beam base girders are employed the bending stress is:

$$f_{bc} = \frac{M_y}{I_{xx}} = \frac{M}{Z_{xx}} = \frac{137.5 \times 10^6}{(1324 \times 10^3)} = 103.8 \text{ N/mm}^2$$

The effective length of the cantilever portion of the girders is taken as 2.5 times the length of the cantilever since it is continuous and restrained on the pier base. It could be argued that a value of $L_e = 1.5L$, table 35, BS 5975, could be used if the ends of the two beams are plated together but this would generally be viewed as an additional safety measure. The ratio L_e/r_y is therefore equal to a maximum of 37.5 and $D/T =$ 25.7 which results in an allowable bending compressive stress of, $p_{bc} = 165\,\mathrm{N/mm^2}$. Therefore these available base beams are more than adequate in bending.

From the shear force in the cantilevered base beam the shear stress

$$f_q = \frac{458 \times 10^3}{(2 \times 357 \times 9.7)} = 66\,\mathrm{N/mm^2}$$

and for an unstiffened 'I' beam web, (BS 5975 and Appendix A) the web shear stress should not exceed $100\,\mathrm{N/mm^2}$. Web stiffeners are not therefore required for shear in the cantilevered section of the base girders and neither are load-bearing stiffeners at other tower support points to prevent web crushing or web buckling. Nevertheless it would be advisable to insert web stiffeners.

The deflection of the cantilevered section is approximated by, $PL^3/(3EI) = 0.3\,\mathrm{mm}$. Normally, to prevent the possibility of load redistribution and to maintain efficient control of geometry the deflection of the critical base girders would be restricted to about a five hundredth of the cantilever span which is 1.2 mm.

A check has to be carried out on the overall stability of the structure to horizontal sliding and to overturning. For this scheme the falsework structure is stable for all feasible load combinations except possibly to horizontal wind forces with no deck cast or materials stored. In this situation, reaction with the piers or holding down bolts to the pier foundations are required. If no assistance from the piers is obtainable then the full wind force across the bridge plus uplift forces on the formwork would have to be resisted by self-weight of the structure and formwork in conjunction with holding down bolts.

The deflections of the Warren girders and braced towers can be found directly using a stiffness analysis which requires computer facilities or very roughly for a heavily braced system the moment of inertia of lattice is approximately equal to the sum of the boom sectional areas times the square of their distances from the neutral axis of the lattice from which the deflection can be approximated using normal beam deflection formulae. More appropriately a flexibility approach assuming a statically determinate pin-jointed system can be used as follows.

Firstly, the girder or braced tower system is analysed by the method of sections or other method to determine the forces in each element for the applied loading. Now where we wish to find the structural deflection we apply unit load in the required direction and analyse the structure to find the element forces induced by the unit load. The required deflection is then given by:

$$\delta = \Sigma[(\text{force in each element from applied loading}) \times (\text{force in each element from the unit load}) \times (\text{length of each element})/(EA \text{ of each element})]$$

Vertical deflection at mid-span of the deck support Warren girders found by this method is 29.7 mm maximum. This calculated deflection is likely to be greater than the in service value for the loaded falsework since joint fixity has been ignored. To allow for the deflection of the falsework structure the girders may be precambered or the formwork packed. This estimated deflection is of order a six hundredth of the span. To avoid the optical illusion of apparent downward bending of road bridge spans they are often given an upward curvature of approximately one hundredth of the span. If this is the case the predicted falsework deflections can be accommodated in the initial falsework placement.

Vertical deflection of the top boom of one of the lattice girders is approximated by:

$$\frac{WL^4}{(384\,EI)} = 0.12\,\text{mm}$$

which is small in comparison with normal construction tolerances.

Sway of a braced row of towers would be resisted by the deck falsework and generally by bracing against the pier strong axes. For maximum lateral forces and if the only resistance available to restrict sway is to be obtained from the braced towers the sway approximated by further application of flexibility methods results in $\Sigma Ppdx/(AE) = 2.54\,\text{mm}$. This is satisfactory since it is within general construction tolerances especially for the falsework height and it is highly unlikely that full steady maximum design wind force would be sustained throughout a major pour.

In the above, general checking of the falsework structure has been carried out. Additional checks should be made regarding the capacity of the pier foundations, the design of the bolted and welded connections, the anchor bolts at the base beam, bending of bearing pads and beam flanges at supports and temperature stresses where this might present a problem. Not least the condition of the construction materials must be carefully inspected and close supervision of erection of the falsework and of the construction work is required.

5.3.2 General scheme 2

The design forces acting on this structure (fig. 5.11) and the assessment of the falsework system are similar to general scheme 1. This scheme is triangulated and allows lighter deck support girders to be used since the bending moment is proportional to the square of the girder span and the deflection to the fourth power of the span.

Lateral bracing across both the towers and the inclined columns resists lateral sway in conjunction with the piers inertia about their strong axes. Level adjustment is obtained by means of telescopic tower and column feet. Pin hinges are employed to angle the inclined columns, eliminate bending of the columns and spread the load evenly to the column elements.

The towers employed in this scheme may be lighter than those in scheme 1 since effectively twice the number of lattice towers carry the deck loads to the bases.

5.3.3 General scheme 3

In this scheme a general scaffolding system is used (fig. 5.12) in conjunction with 'I' beams, which form an open span over the river and bases for the verticals. Assessment of this design is carried out in easy stages.

The vertical loading is approximately the same as for scheme 1 but the lateral wind load is greater because of the forest of scaffolding. The vertical load-bearing capacity of the system is found by checking the capacity of a single scaffold pole. The pole chosen for this check is a vertical and the portion investigated is that between the foot ties and the next layer of lateral bracing. For example the maximum compressive stress in element $1-2$ is:

$$f_c = \frac{\text{load}}{\text{tot. cross-sectional area of vert. poles}} = \frac{5262 \times 10^3}{(557 \times 209)} = 45.2\,\text{N/mm}^2$$

The effective length of a vertical between bracing is generally $0.7L$ but at their bases they are not continuous, although tied about each axis, and therefore the appropriate effective length for the load capacity check is $0.85L$ to $1.0L$. This system employs normal scaffold tubes and couplers therefore an effective length of $1.0L$ is taken to allow for any contingencies. The slenderness ratio, $L_e/r = 1500/15.7 = 95.5$ which yields a value of $p_c = 75.4\,\text{N/mm}^2$ for new tubes or $64.1\,\text{N/mm}^2$ for used tubes, these

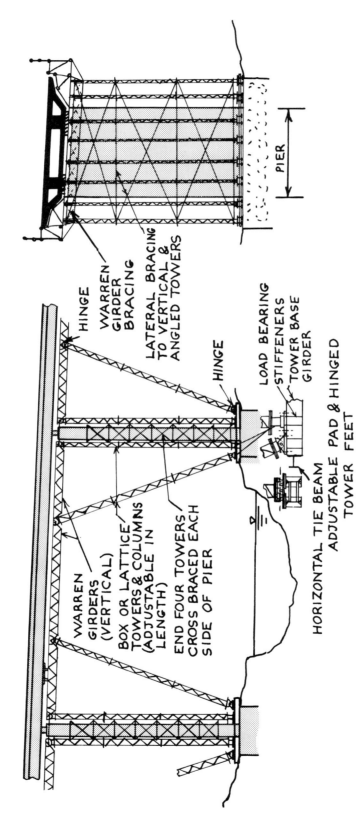

HINGE

WARREN GIRDER BRACING

LATERAL BRACING TO VERTICAL & ANGLED TOWERS

HINGE

LOAD BEARING STIFFENERS

TOWER BASE GIRDER

ADJUSTABLE PAD & HINGED TOWER FEET

HORIZONTAL TIE BEAM

END FOUR TOWERS CROSS BRACED EACH SIDE OF PIER

BOX OR LATTICE TOWERS & COLUMNS (ADJUSTABLE IN LENGTH)

WARREN GIRDERS (VERTICAL)

PIER

Figure 5.11 General scheme 2 for bridge falsework

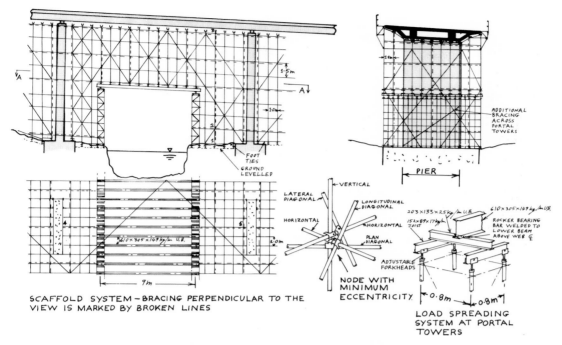

Figure 5.12 General scheme 3 for bridge falsework

values being derived from table 23, BS 5975. The p_c value is lower than that for normal grade 43 steel sections to allow for general eccentricities of couplers.

The lateral stability can be assessed by examination of the capacity of the cross bracing and the forces acting on the width between bracing (three verticals if the portal towers are braced to take the lateral forces on the portal span). For example on bracing 1–2–3 the maximum lateral force is $405/6 = 67.5$ kN maximum and for four bracing diagonals acting the compressive stress in a diagonal is given by:

$$f_c = \frac{67.5 \times 10^3}{[1/\sqrt{(1 + 2.25)}] \times 557 \times 14} = 54.6 \text{ N/mm}^2$$

Taking $L_e = 1.0L$ for a diagonal between couplers the ratio $L_e/r = 1802/15.7 = 114.8$ therefore from table 23, BS 5975 the allowable compressive stress is approximately 60 N/mm² for new tubes and 50 N/mm² for used tubes. It is therefore evident that close supervision and the use of new tubes is required if extra diagonals are not to be included. Some additional vertical loading also acts on the system from this source and can be approximated by normal bending approximations and employing a global value for the second moment of a row of vertical poles about the falsework centre line (e.g. $I_{xx} = \Sigma A y^2$) and obtaining the maximum force in a vertical from the tributary portion of the stress diagram relating to linear bending stress, M_y/I_{xx} (fig. 5.13).

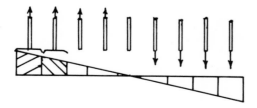

Figure 5.13 Approximate bending force transfer to poles in a vertical row

Longitudinal stability is similarly assessed by examination of the capacity of bracing on line 4–5 (fig. 5.12) and the forces which act on it. The maximum compressive stress on a diagonal is given by:

$$f_c = \frac{215 \times 10^3}{2 \times 4 \times 557} = 48.3 \text{ N/mm}^2$$

The ratio $L_e/r = 1.0L/r = 1802/15.7 = 114.8$ and from table 23, BS 5975, $p_c \simeq 60$ N/mm² for new tubes and $p_c \simeq 50$ N/mm² for used tubes, therefore the bracing scheme is suitable. The lateral stability is further increased by the plan bracing which also serves to eliminate differential lateral movement between vertical rows of scaffolding.

The portal spanning the river comprises towers fabricated from scaffold poles, which can be assessed as for the towers in scheme 1, and simply supported 'I' beams, which are assessed by computing the bending compressive stress from the maximum bending moment and comparing this stress with the allowable stress as for scheme 1. Checks have to be made in case any load-bearing stiffeners are required in any of the 'I' beams. The maximum bending moment is

$$\frac{wL^2}{8} = \frac{WL}{8} = \frac{3046 \times 9}{11 \times 8} = 311.5 \text{ kNm}$$

The maximum bending stress corresponding to this moment is given by:

$$\frac{M_y}{I_{xx}} = \frac{M}{Z_{xx}} = \frac{311.5 \times 10^6 \times 305}{1.24 \times 10^9} = 76.6 \text{ N/mm}^2$$

The ratio L_e/r_y is taken as $1.2(L + 2D)/r_y$, using the data in table 34, BS 5975, since full end restraint of the girders is not guaranteed. Therefore $L_e/r_y = 1.2(9 \times 10^3 + 2 \times 610)/69.9 = 175.5$ and for $D/T = 30.9$ the allowable value of compressive bending stress is, from table 21, BS 5975, 82 N/mm² such that the 610 × 305 × 149 kg/m Universal Beams are adequate for the portal span.

The bending moment in the upper cross beam of the tower capping load spreading arrangement is $WL/4 = 138 \times 0.8$ max$/4 = 27.6$ kNm maximum. If 203 × 133 × 25 kg/m Universal Beams are used then the bending stress, $f_{bc} = 119$ N/mm². $D/T = 25.9$ and $L_e/r_y = 1.2L/r_y = 1.2 \times 800/31 = 31$ since some restraint in the span is obtained from the portal beams. Using these values in table 21, BS 5975, we obtain $P_{bc} = 165$ N/mm². The bending moment in a lower spreader beam is 13.8 kNm and $f_{bc} = 119$ N/mm². A check has also to be made in case web stiffeners are required.

The maximum vertical stress in a tower leg, excluding any force resulting from lateral loads, is 61.1 N/mm². For a leg the maximum value of the slenderness ratio is, $L_e/r = 1.0 \times 750/15.7 = 47.8$ therefore $p_c = 113$ N/mm² for new tubes and 96.2 N/mm² for used tubes.

In schemes 1 and 2 the falsework was founded upon the pier bases and therefore appraisal of soil conditions and base pads was unnecessary. In this scheme the pole and tower footings are mainly directly onto the soil although some rest on the pier foundations thus emphasising the need to restrict differential settlement. To provide adequate footings for those poles not resting on the pier foundations the ground requires excavation (a minimum of 0.5 m) to virgin soil and the provision of ground beams, base pads or a base slab. Piles may be required to support any of these in the case of weak strata. Great care must be exercised where falsework is founded on fill materials. In this scheme the soil was tested and found to be consistently stiff clay but to minimise any differential vertical movements economically and to provide sufficient horizontal restraint to counteract the lateral forces on the system a concrete screed was used in conjunction with an anchored steel grillage base for the falsework.

It is vitally important in the erection of this scheme to ensure that the verticals are plumbed, forkheads have no eccentricity and supports are at right angles to struts, the eccentricities of nodal connections are kept to a minimum and that the load from each lintel beam is spread equally to the tower legs.

5.3.4 General schemes 4a and 4b

Similar construction problems are associated with schemes 4a and 4b (fig. 5.14). Among the points to note are that jacking of the rig legs should only be executed immediately above bridge supports or structural towers specifically designed for the purpose, to anchor the rig against lateral or longitudinal movement or overturning and to avoid inducing unnecessary bending in the deck under construction. In 4a the concrete has to be designed to give the desired strength and rate of curing required to allow maximum construction rate to be achieved otherwise an efficient crew may work themselves ahead of the concrete curing process. In 4b the prestressing cables should be grouted in immediately after tensioning so that if overstress of the deck should occur then the strain in the cables would be concentrated and control of cracking would be obtained, which in turn increases the ultimate capacity of the system.

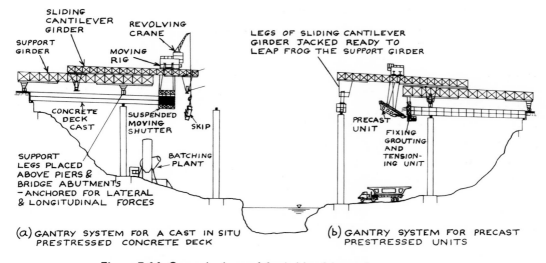

(a) GANTRY SYSTEM FOR A CAST IN SITU PRESTRESSED CONCRETE DECK

(b) GANTRY SYSTEM FOR PRECAST PRESTRESSED UNITS

Figure 5.14 General scheme 4 for bridge falsework

Chapter 6
Design Checks and Safety Aspects for Falsework and Access Scaffolding

6.1 GENERAL

Many design points and safety aspects which are often considered to be trivial or merely routine as well as the more intricate or complex design problems are worthy of consideration in this section. This is borne out in the sections below in which a number of common practices and problem areas are examined.

6.2 COMMON PROBLEM AREAS

Inaccurate estimation of the magnitudes and combined actions of probable loadings represents an extremely common source of difficulty in falsework design and is often the root cause of failures attributed to other causes. Some guidance on force actions is given in section 5.2.

Many aspects feature in the choice of a falsework system. Important among these are economics of the situation, whether to hire or buy, speed of erection, ease of transportation, durability, complexity and versatility. In cases where used systems are employed inspection has to be made for rust cavitation, cracked elements, weld fatigue, bent, crushed or buckled sections and generally damaged units, all of which would reduce the load-carrying capacity of the scheme. Often, reduced loadings are recommended for used systems (BS 5975, table 23).

Site supervision of the erection (and dismantling) of falsework and ancillary equipment is highly important to ensure correct fabrication, component use, foundation provision, plumbing of towers and verticals and levelling of structures.

Wind loads are often as significant as vertical loads on falsework since the structures are generally primarily designed to withstand vertical loading. Special care should be exercised regarding wind loading on freestanding falsework and access scaffolds especially in the absence of vertical loading or where there are eccentric vertical force actions.

It is often wise to anticipate possible damage from site traffic or machinery and to provide crash barriers at vulnerable points, protective frames and safety nets for personnel.

Care should be exercised in the use of guyed falsework in case the guys resonate or gallop in wind or stretch and proper foundations for guys may be difficult to construct.

Permanent shuttering and proprietary units should be used strictly in accordance with the manufacturer's specifications unless adequate analysis or testing of an alternative use is carried out.

Vibrations from construction equipment can be a problem regarding connection fixity and the movement of sloping supports. Provision can be made to eliminate problems arising from this source by securing all supports and employing bolted connections incorporating spring washers or lock nuts.

Foundation checks should be carried out in all cases and often it will be necessary to provide a grillage even where the permanent structure foundations are available for a proportion of the falsework structure.

In addition to the provision of properly designed members and connections the overall stability of a falsework system has to be examined. This is important not only in large systems but equally so in access scaffolds and portable inspection and maintenance towers. For instance in portable towers internal ladders should be provided since in many cases such structures can be overturned by a man climbing up the outside bracing (fig. 6.1g).

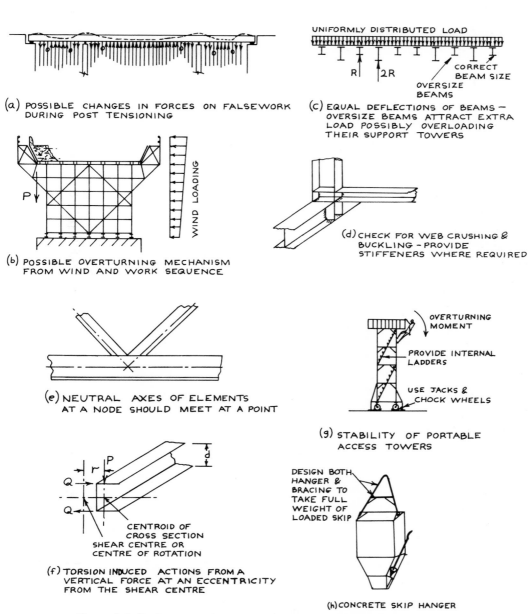

(a) POSSIBLE CHANGES IN FORCES ON FALSEWORK DURING POST TENSIONING

(b) POSSIBLE OVERTURNING MECHANISM FROM WIND AND WORK SEQUENCE

(c) EQUAL DEFLECTIONS OF BEAMS — OVERSIZE BEAMS ATTRACT EXTRA LOAD POSSIBLY OVERLOADING THEIR SUPPORT TOWERS

(d) CHECK FOR WEB CRUSHING & BUCKLING – PROVIDE STIFFENERS WHERE REQUIRED

(e) NEUTRAL AXES OF ELEMENTS AT A NODE SHOULD MEET AT A POINT

(f) TORSION INDUCED ACTIONS FROM A VERTICAL FORCE AT AN ECCENTRICITY FROM THE SHEAR CENTRE

(g) STABILITY OF PORTABLE ACCESS TOWERS

(h) CONCRETE SKIP HANGER

Figure 6.1 Design and safety aspects of construction and access systems

The sequence of construction is often important in falsework design to ensure stability and restrict differential deflections which could become cast into a finished structure (fig. 6.1b).

Where a falsework design specifies equal 'I' beams to provide a deck support system then great care should be exercised if a mixture of oversize and specified size beams are employed on an availability basis. In such an instance the oversize beams, with greater bending stiffness, will attract greater loads and increase the loadings on their support towers. For instance, if equally spaced parallel beams support a uniformly distributed loading and every other beam has twice the moment of inertia in bending as the others then these beams attract twice the load of the others for equal deflected forms of all beams (fig. 6.1c).

At nodes of scaffold structures and trusses the neutral axes of the various members should meet at a point or at a minimum eccentricity to avoid additional stresses being set up (fig. 6.1e).

There have been several recent collapses of falsework used to support structures to be post-tensioned. The post-tensioning process usually results in changes of level at various positions and a redistribution of loading on the falsework system. An analysis must be carried out to predetermine level changes and the falsework designed to accommodate the movements such that neither the structure nor the falsework are overstressed (fig. 6.1a).

Struts should be checked for buckling about both axes and the minimum net cross-sectional area of ties is appropriate for the assessment of their load-carrying capacity. The effective length of a strut may be different about each axis and a guide to the appropriate effective length is given in BS 449 and BS 5975.

At all support points the webs of 'I' beams should be checked for buckling and crushing and stiffeners provided where required (fig. 6.1d).

The effective length of beams largely governs their capacity to resist lateral buckling. The effective length can be assessed by means of the guides in BS 449 or BS 5975. The effective lengths can be controlled by the provision of adequate structural bracing between elements.

An important source of secondary stresses in some strcutures is torsion which can be inadvertantly induced in a structure by incorrect positioning of loading. For example, the force acting on the channel in fig. 6.1f through the centre of gravity of the cross section creates a torque about its shear centre of Pr which in turn induces a couple Qd which results in both shear and bending of the channel flanges in addition to the vertical shear and bending from the direct action of the load P. In general terms eccentrically positioned vertical loading on a strut induces bending in addition to compression and a check must be made on the capacity of such members to resist the combined action. Eccentric loading at right angles to a beam causes combined bending and torsion and a check must be made on capacity for such actions.

When slipforming tall structures gravity twist can be a problem and plumbing by means of optical instruments is often difficult. Continuous checks on alignment must be made and a plumb line should be used where required.

When excavating within diaphragm walls, normal foundations, shafts or trenches provision has to be made for the lateral pressure on the walls or piling. In many cases this can be counteracted by internal bracing but where this is not practical because of construction processes or other restrictions lateral restraint can be provided by means of anchor rods to piles or diaphragm wall section anchor blocks situated outside the excavation zone.

If a concrete skip is provided with bracing across its hanger then the bracing should be designed to take the full load since the crane hook may be attached to the bracing to reduce the clearance height (fig. 6.1h). If the bracing is not sufficiently strong then failure may occur with the probability of injury to those operating the pour from the skip.

Propping and repropping of floors can cause overstressing of the supporting floors in multi-storey construction if not carefully planned. Random repropping may also result in reverse bending in green units which will crack and may be permanently weakened.

The above should not be viewed as an exhaustive list of design and safety points, since space does not allow for all contingencies to be expanded upon but some common weaknesses and potentially unsafe practices have been highlighted.

6.3   SIMPLE FALSEWORK CHECK PROCEDURE

In section 6.2 a few of the many points were discussed which should be considered in the design, erection and use of falsework. In general the overall checking of a falsework system can be carried out as indicated in table 6.1 with the following items considered in detail:

(1) Check vertical, lateral and longitudinal dead, superimposed, wind, ice, snow and other relevant loadings (see section 5.2).

(2) Check formwork for strength, force components, deflections and centering.

(3) Check the vertical capacity of the falsework, its resistance to instability from lateral loading and to overturning.

(4) Check the design of all connections, nodes and beams for strength, eccentricity, buckling and crushing and all bolts and welds. Great care should be taken in the re-use of materials, bolts, which may have thread damage and be of unknown quality, and other fixings and especially in the use of welded connections which may be in shear or tension.

(5) Check the deflections of the overall falsework system and components.

(6) Check that load distribution at transfer points is achieved in an appropriate manner.

(7) Check the adequacy of the foundations for vertical, overturning and shear forces.

(8) Check that erection of the falsework can be carried out safely and efficiently.

(9) Provide adequate warning against alteration of the falsework scheme and provide proper drawings showing all members and details.

(10) Provide a comprehensive guide as to the exact erection and operation procedures to be adopted regarding the falsework use.

(11) Before the design is finalised recheck the loadings. If the loadings are higher than the initial estimates or if the load combinations result in reduced stability of a falsework system then recheck the design and alter as necessary.

In many instances a number of falsework schemes will be feasible. Effort must not be spared in assessing individual schemes at initial stages if a wrong choice of system is to be avoided.

Table 6.1 Flow path for falsework checking

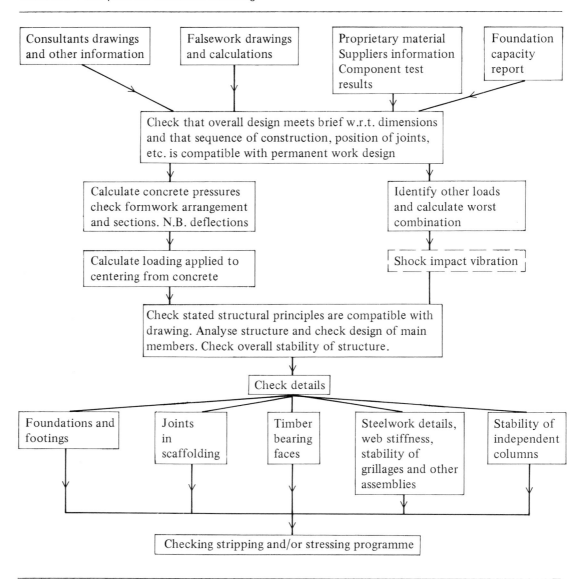

Chapter 7
References

1. British Standard 5975 (1982). 'Code of practice for falsework', British Standards Institution.
2. British Standard CP 3: Chapter V: Part 1 (1967). 'Dead and imposed loads', British Standards Institution.
3. British Standard CP 3: Chapter V: Part 2 (1972). 'Wind loads', British Standards Institution.
4. British Standard 449: Part 2 (1969). 'The use of structural steel in building', British Standards Institution, 1969 and later amendments.
5. *Steel Designers' Manual* (1979). 4th edn., Crosby Lockwood Staples, London, Granada Publishing Ltd.
6. Irwin, A. W. (1983). 'Design of tall buildings of shear wall form', Construction Industry Research and Information Association.
7. Proceedings of CIRIA Conference (1981). 'Wind engineering in the eighties', Construction Industry Research and Information Association.
8. Irwin, A. W. (1973). 'Critical load selection for economic analysis of unusually shaped buildings', *Build International*, Applied Science Publishers, Vol. 6, 387–403.
9. Irwin, A. W. and Jeary, A. P. (1976). 'Vibrations of a nuclear power station charge hall', *Earthquake Engineering and Structural Dynamics*, Vol. 4, 221–229.
10. Benjamin, J. R. and Cornell, C. E. (1970). *Probability, Statistics and Decision for Civil Engineers*, McGraw-Hill Book Co.
11. British Standard 648 (1964). 'Schedule of weights of building materials', British Standards Institution.
12. Irwin, A. W. (1981). 'Live load and dynamic response of the extendable front bays of the Playhouse Theatre during "The Who" concert', *Report for Lothian Region Architecture Department*.
13. International Conference of Building Officials: Uniform Building Code (1960–1980) various editions, Pasenda, USA.
14. Rubinstein, M. F. (1970). *Structural Systems – Statics, Dynamics and Stability*, Prentice-Hall, Inc.
15. Formwork – Concrete Society Technical Report No. 13 (1977).
16. Falsework – Concrete Society Technical Report TRCS 4 (1971).
17. British Standards Institution, Draft Code of Practice for Falsework (1975).
18. Association of Finnish Plywood Industry FPDA technical publication No. 20 (1975).
19. Council of Forest Industries of British Columbia, Design of Plywood Formwork (1979).

20. HMSO, Final Report of the Advisory Committee on Falsework.
21. British Standard 4978 (1973). 'Timber grades for structural use', British Standards Institution.
22. British Standard CP 112 (1971). 'The structural use of timber', Part 2, 'Metric units', British Standards Institution.
23. British Standard 4074 (1966). 'Metal props and struts', British Standards Institution.
24. British Standard 1139 (1964). 'Metal scaffolding', British Standards Institution.

Appendix A
Data for Formwork Design

A.1   LOAD/SPAN GRAPHS FOR TIMBER SECTIONS

Basic graphs Nos. 1 – 7 (fig. A.3) give the safe load/span relationship for standard timber sections used in falsework. Appropriate permissible stresses for the loading conditions in Table A are used (see table A.1), i.e. simply supported single span; total U.D.L. = W; no load sharing; strength class SC3 timber; actual dry timber dimensions.
   In the general case the limiting value of safe load is determined by:

        (i) shear stress
or   (ii) bending stress
or (iii) deflection.

   For spans exceeding 1 m the deflection limit of 3 mm will apply. For spans less than 1 m the deflection limit of 0.003 × span will apply.

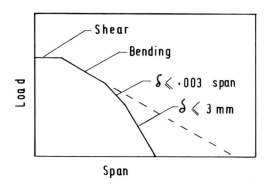

Figure A.1 Safe load line

   Figure A.1 shows how the safe load line is constructed using the limiting criteria. If, for any reason the deflection limit is not applicable the bending limit line may be extended forward.
   The graphs may also be used, by the application of the coefficients in table A.1 to basic safe load values, for a range of loading specifications;

   (i) other strength class timber
   (ii) equal span continuous beam U.D. loading
   (iii) load sharing.

Coefficients to cover the loading conditions in 2.5.2 could be calculated and applied in the same way. However, for a one-off design involving non-standard loading it will be more convenient to obtain timber sections and spacing from first principles.

Table A.1

		USE ACTUAL SIZE OF DRY TIMBER			SAFELOAD = BASIC GRAPH VALUES X COEFFICIENT SELECT MINIMUM LIMITING CRITERIA					
LOADING TABLE	TIMBER	PERMISSIBLE STRESSES		E VALUE	SIMPLY SUPPORTED			CONTINUOUS		
	STRENGTH CLASS	SHEAR N/MM²	BENDING N/MM²	1 N/MM²	SHEAR	BENDING	DEFLEC-TION	SHEAR	BENDING	DEFLEC-TION
A	Sc3	1.34	6.05	4838	BASIC GRAPH VALUES			.83	1.25	1.89
B	Sc4	1.34	8.67	5616	1.00	1.43	1.16	.83	1.79	2.19
C	Sc5	1.75	11.87	6134	1.31	1.96	1.27	1.09	2.45	2.39
D	Sc3	1.47	6.66	7344	1.10	1.10	1.52	1.27	1.38	2.87
E	Sc4	1.47	9.54	8035	1.10	1.58	1.66	1.38	1.97	3.13
F	Sc5	1.92	13.06	9418	1.43	2.16	1.95	1.62	2.7	3.68

(Rows A, B, C grouped under **NO LOAD SHARING**; rows D, E, F grouped under **LOAD SHARING**)

Barri Load = 16.86 × 1.25' = 20.7

Alternative loading conditions.

(i) Simply supported

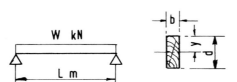

$$\text{max. shear stress} = \frac{1.5 \times 0.5W}{bd}$$

$$\text{max. bending moment, } M = \frac{WL}{8}$$

$$\text{max. bending stress} = \frac{My}{I}$$

$$\text{deflection, } \delta = \frac{5WL^3}{384EI}, \text{ least of 3 mm or } 0.003 \times \text{span}$$

(ii) Continuous over three or more spans

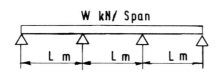

$$\text{max. shear stress} = \frac{1.5 \times 0.6W}{bd}$$

$$\text{max. bending moment, } M = \frac{WL}{10}$$

$$\text{deflection, } \delta = \frac{1.0WL^3}{145EI}, \text{ least of 3 mm or } 0.003 \times \text{span}$$

Load sharing

If members are spaced not more than 600 mm apart, and there is adequate provision for lateral distribution of loads by means of decking or joists spanning at least three supports, stresses may be increased by a factor of 1.1.

For calculation of deflection of load-sharing systems the mean value for the modulus of elasticity is used.

Examples

1. What is the safe U.D.L. per span on a SC3 150 × 150 continuous beam with supports at 1.4 m centres, no load sharing?

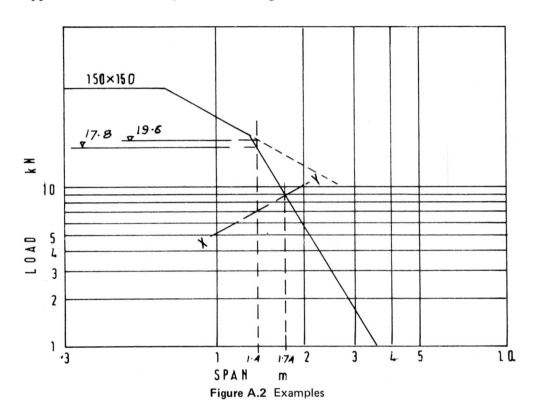

Figure A.2 Examples

From safe load graph No. 7 (fig. A.3) and table A.1:

Basic load bending = 19.6 kN, design load = 19.6 × 1.25 = 24.5 kN
          deflection = 17.8 kN, design load = 17.8 × 1.89 = 33.04 kN

Bending stress is limiting criteria. Safe U.D.L. = 24.5 kN.

2. Find maximum spacing of supports to a SC4 150 × 150 primary beam in a soffit centering arrangement, continuity and load sharing will apply.

Loading from secondary beams may be taken as U.D.L. of 16 kN/m.
From table A.1, loading table E, coefficient for deflection = 3.13
                          and for bending = 1.97
Convert design load to basic load:

If deflection is critical basic load $= \dfrac{16}{3.13} = 5.11\,\text{kN/m}$

If bending is critical basic load $= \dfrac{16}{1.97} = 8.12\,\text{kN/m}$

Deflection is critical. Basic load for span of 2 m $= 10.22\,\text{kN}$.
Construct $X - Y$, gives span $= 1.74\,\text{m}$ (see fig. A.2).

Check   Design U.D.L. $= \dfrac{8.9 \times 3.13}{1.74} = 16.00\,\text{kN/m} - \text{O.K.}$

fig A3

Graph I

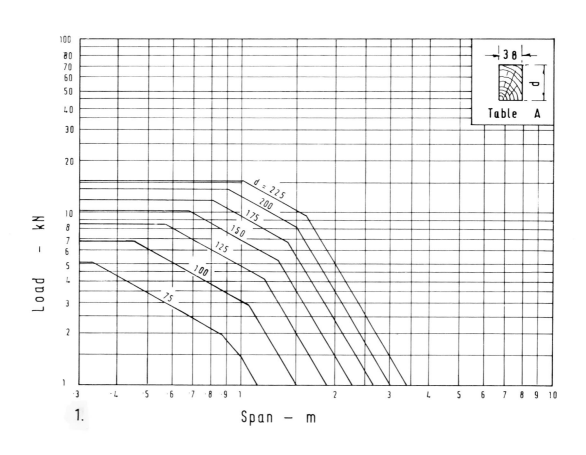

1.

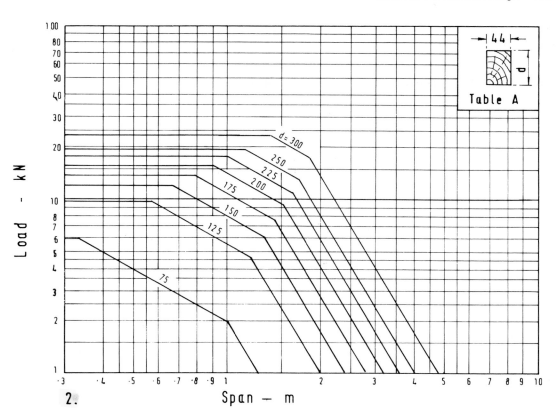

2.

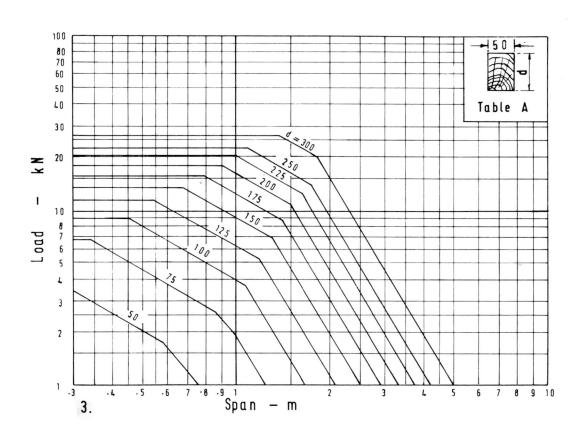

3.

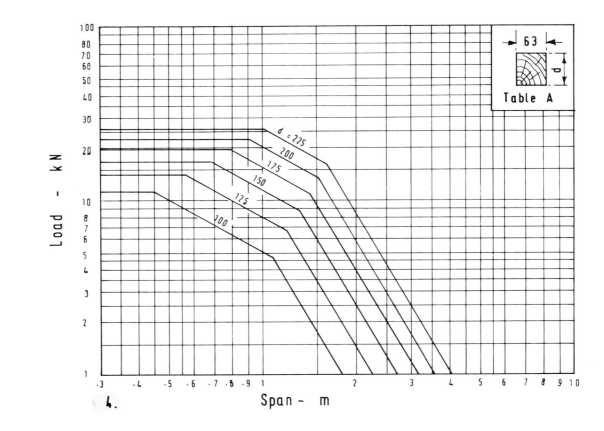

Graph IV

4.

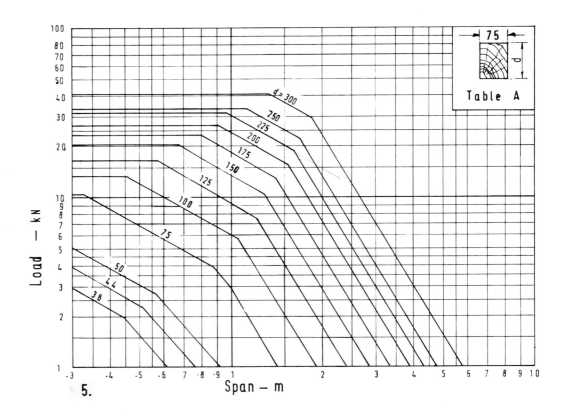

Graph V

See p. 67.
16.56 Cond.

5.

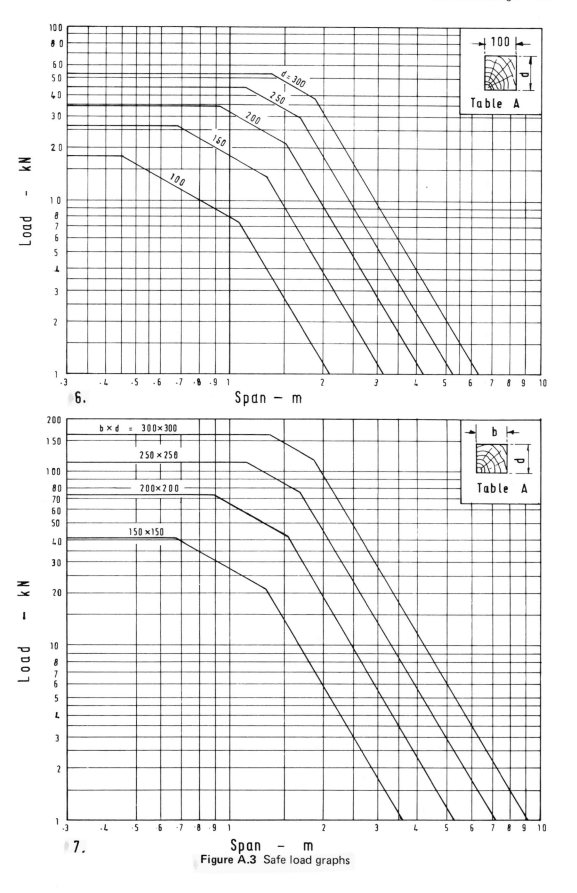

Figure A.3 Safe load graphs

A.2 FINNISH BIRCH-FACED PLYWOOD – GRAPH RELATING CONCRETE PRESSURE, PLYWOOD THICKNESS AND STUD SPACING[18]

Figure A.4 relates concrete pressures to plywood thicknesses for a range of deflection to span ratios at moisture conditions suitable for formwork use.

The dotted lines superimposed on the graph are deflection to span ratio limitations imposed by the allowable bending stresses.

This graph relates only to standard Finnish plywood used with the grain at right angles to supports. It is based on the allowable working stresses in table A.2 on the assumption that the plywood is continuous over several spans.

Table A.2 Allowable working stresses, elastic constants and section properties

Allowable working stresses									
Bending									
‖ to face grain	(N/mm²)	12.26	11.76	11.07	10.50	9.93	9.37	8.83	8.27
⊥ to face grain	(N/mm²)	5.67	7.08	7.57	7.83	7.80	7.40	7.12	6.33
Rolling Shear (bending flat)									
‖ to face grain	(N/mm²)	0.64	0.80	0.91	1.01	1.10	1.10	1.05	1.04
⊥ to face grain	(N/mm²)	0.45	0.49	0.63	0.65	0.62	0.59	0.51	0.49
Elastic Constants									
Bending									
‖ to face grain	(kN/mm²)	8.81	8.65	8.45	7.26	6.90	6.55	6.19	5.90
⊥ to face grain	(kN/mm²)	2.48	3.11	3.23	3.29	3.29	3.58	3.67	3.65
Section Properties (per metre width)									
Area	(mm² × 10³)	6.5	9.3	12.0	14.8	17.6	20.4	23.2	26.0
Section modulus	(mm³ × 10³)	7.05	14.4	24.0	36.4	51.5	69.2	89.9	112
Second moment of area	(mm⁴ × 10³)	22.9	66.8	143	270	456	709	1040	1465

Note – The data have been derived experimentally and allow for the wet conditions met in concreting (up to 27% moisture content).

The allowable working stresses incorporate an overall safety factor of three based on characteristic stresses and long term loading.

Procedure

1. Determine concrete pressure as in section 2.4 or otherwise and locate this pressure on the concrete *pressure axis.*
2. Move vertically to intersect the *plywood thickness* curves. These points of intersection show possible plywood thicknesses for that pressure. Choose an initial thickness to examine.
3. Note the position of the chosen point in relation to the dotted lines which show *deflection/span ratio limitations* imposed by the allowable bending stresses. By interpolation between the dotted lines estimate the *deflection/span limitation* of that point and this will be the maximum value of deflection/span allowable. The *deflection/span line* chosen on the left-hand side of the graph should not exceed this value (note 1/250 exceeds 1/300).
4. Move horizontally from the chosen point to intersect the required *deflection/span line* having regard to the limitations imposed in paragraph 3 above.
5. Move vertically down from this intersection to determine the *stud spacings* given for the chosen conditions and thickness.
6. Repeat the procedure for the other possible thicknesses indicated in stage 2, if this is considered necessary.

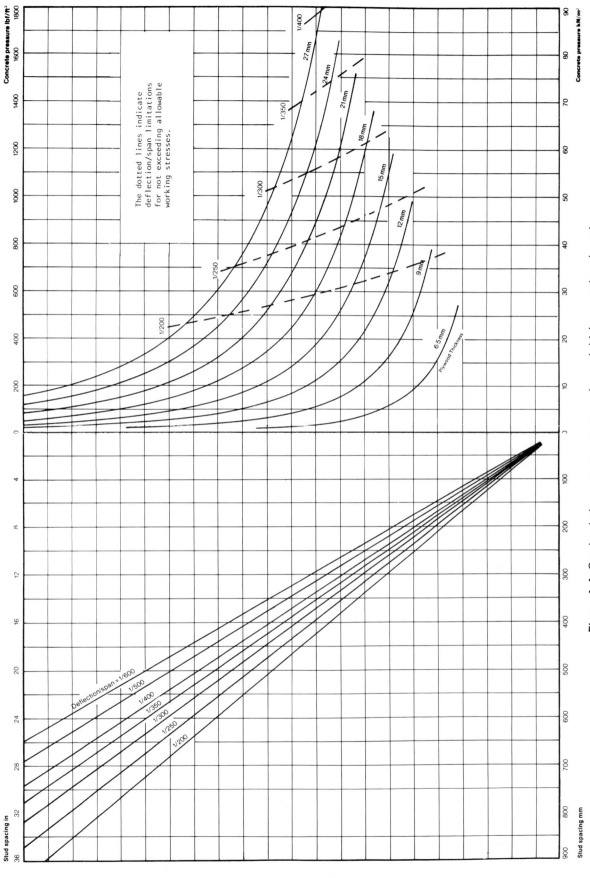

Figure A.4 Graph relating concrete pressure, plywood thickness and stud spacing

A.3   LOAD/SPAN TABLES FOR FORMWORK DESIGN WITH COFI* EXTERIOR DOUGLAS FIR PLYWOOD

Table A.3 gives allowable loads for two thicknesses of COFIFORM plywood. For design information on other grades of formwork plywood consult reference 19.

Table A.3

Allowable loads for COFIFORM – Unsanded Grades

Width of Supports	Span from centre to centre of Supports	Face Grain Perpendicular to Supports					
		20·5mm – 7 ply			18·5mm – 7 ply		
		Loads governed by			Loads governed by		
		Flexure or Shear	Deflection*		Flexure or Shear	Deflection*	
			//270	//360		//270	//360
mm	mm	kN/m²	kN/m²	kN/m²	kN/m²	kN/m²	kN/m²
	200	65·1	164·6	123·4	58·0	137·4	103·0
	250	50·6	104·4	78·3	45·1	86·4	64·8
	300	41·4	70·8	53·1	36·9	58·1	43·6
	350	35·0	50·2	37·6	31·2	40·8	30·6
	400	29·4	36·8	27·6	25·3	29·8	22·3
25	450	23·6	27·7	20·8	20·0	22·3	16·7
to	500	19·1	21·3	16·0	16·2	17·1	12·8
50	550	15·8	16·7	12·6	13·4	13·4	10·0
	600	13·3	13·4	10·0	11·2	10·6	8·0
	650	11·3	10·8	8·1	9·6	8·6	6·4
	700	9·7	8·9	6·7	8·3	7·0	5·3
	750	8·5	7·4	5·5	7·2	5·8	4·4
	800	7·4	6·2	4·6	6·3	4·9	3·7

*Deflection limited to 1/270th or 1/360th of span

Allowable loads for COFIFORM – Unsanded Grades

Width of Supports	Span from centre to centre of Supports	Face Grain Perpendicular to Supports					
		20·5mm – 7 ply			18·5mm – 7 ply		
		Loads governed by			Loads governed by		
		Flexure or Shear	Deflection*		Flexure or Shear	Deflection*	
			//270	//360		//270	//360
mm	mm	kN/m²	kN/m²	kN/m²	kN/m²	kN/m²	kN/m²
	200	75·9	251·4	188·6	67·6	210·3	157·7
	250	56·9	144·3	108·2	50·7	119·9	89·9
	300	45·5	92·3	69·2	40·6	76·1	57·1
	350	37·5	63·0	47·3	33·1	51·5	38·6
	400	29·8	44·9	33·7	25·3	36·4	27·3
51	450	23·6	33·1	24·8	20·0	26·7	20·0
and	500	19·1	25·0	18·8	16·2	20·1	15·1
wider	550	15·8	19·4	14·5	13·4	15·5	11·6
	600	13·3	15·3	11·5	11·2	12·2	9·2
	650	11·3	12·2	9·2	9·6	9·7	7·3
	700	9·7	10·0	7·5	8·3	7·9	5·9
	750	8·5	8·2	6·2	7·2	6·5	4·9
	800	7·4	6·9	5·2	6·3	5·4	4·1

*Deflection limited to 1/270th or 1/360th of span

Notes on table A.3 - Derivation of information and design assumptions

1. Grade stresses calculated from basic Canadian test data and modified to bring them to a UK long duration 'base', i.e. flexure 12.9 N/mm², shear 0.403 N/mm². Moisture content factors of 0.8 and 0.85 then applied to flexure and shear stress respectively to take into account the wet service conditions. Both stresses then multiplied by a load duration factor of 1.33 to allow for the short-term nature.

*Council of Forest Industries of British Columbia

2. Flexure formula is:

$$w = \frac{F_b \times KS}{0.1071l^2}$$

where w = permissible pressure
F_b = permissible flexural stress
KS = effective section modulus of plywood
l = centre to centre spacing of supports.

Plywood is continuous over at least three equal spans with face grain at right angles to supports.
3. Shear formula is:

$$w = \frac{F_s \times I}{0.625lQ}$$

where w = permissible pressure
F_s = permissible shear stress
I = second moment of area
l = clear span
Q = first moment of area.

Plywood is continuous over at least two equal spans.
4. Total deflection calculated by finite element analysis using elastic constants derived from testing, assuming plywood continuous over at least three spans.
5. Plywood used with face grain at right angles to supports.
6. Plywood used in wet service conditions where its moisture content is likely to be between 22% and 29%.

A.4 GEOMETRICAL PROPERTIES OF TIMBER SECTIONS (ACTUAL DIMENSIONS)

Table A.4

Basic size	Area	Section modulus		Second moment of area		Radius of gyration	
		About x–x	About y–y	About x–x	About y–y	About x–x	About y–y
mm	10^3 mm²	10^3 mm³	10^3 mm³	10^6 mm⁴	10^6 mm⁴	mm	mm
16 × 75	1·20	15·0	3·20	0·562	0·0256	21·7	4·62
16 × 100	1·60	26·7	4·27	1·33	0·0341	28·9	4·62
16 × 125	2·00	41·7	5·33	2·60	0·0427	36·1	4·62
16 × 150	2·40	60·0	6·40	4·50	0·0512	43·3	4·62
19 × 75	1·42	17·8	4·51	0·668	0·0429	21·7	5·48
19 × 100	1·90	31·7	6·02	1·58	0·0572	28·9	5·48
19 × 125	2·38	49·5	7·52	3·09	0·0714	36·1	5·48
19 × 150	2·85	71·2	9·02	5·34	0·0857	43·3	5·48
22 × 75	1·65	20·6	6·05	0·773	0·0666	21·7	6·35
22 × 100	2·20	36·7	8·07	1·83	0·0887	28·9	6·35
22 × 125	2·75	57·3	10·1	3·58	0·111	36·1	6·35
22 × 150	3·30	82·5	12·1	6·19	0·133	43·3	6·35
25 × 75	1·88	23·4	7·81	0·879	0·0977	21·7	7·22
25 × 100	2·50	41·7	10·4	2·08	0·130	28·9	7·22
25 × 125	3·12	65·1	13·0	4·07	0·163	36·1	7·22
25 × 150	3·75	93·8	15·6	7·03	0·195	43·3	7·22
25 × 175	4·38	128	18·2	11·2	0·228	50·5	7·22
25 × 200	5·00	167	20·8	16·7	0·260	57·7	7·22
25 × 225	5·62	211	23·4	23·7	0·293	65·0	7·22
25 × 250	6·25	260	26·0	32·6	0·326	72·2	7·22
25 × 300	7·50	375	31·2	56·2	0·391	86·6	7·22
32 × 75	2·40	30·0	12·8	1·12	0·205	21·7	9·24
32 × 100	3·20	53·3	17·1	2·67	0·273	28·9	9·24
32 × 125	4·00	83·3	21·3	5·21	0·341	36·1	9·24
32 × 150	4·80	120	25·6	9·00	0·410	43·3	9·24
32 × 175	5·60	163	29·9	14·3	0·478	50·5	9·24
32 × 200	6·40	213	34·1	21·3	0·546	57·7	9·24
32 × 225	7·20	270	38·4	30·4	0·614	65·0	9·24
32 × 250	8·00	333	42·7	41·7	0·683	72·2	9·24
32 × 300	9·60	480	51·2	72·0	0·819	86·6	9·24
36 × 75	2·70	33·8	16·2	1·27	0·292	21·7	10·4
36 × 100	3·60	60·0	21·6	3·00	0·389	28·9	10·4
36 × 125	4·50	93·8	27·0	5·86	0·486	36·1	10·4
36 × 150	5·40	135	32·4	10·1	0·583	43·3	10·4
38 × 75	2·85	35·6	18·0	1·34	0·343	21·7	11·0
38 × 100	3·80	63·3	24·1	3·17	0·457	28·9	11·0
38 × 125	4·75	99·0	30·1	6·18	0·572	36·1	11·0
38 × 150	5·70	142	36·1	10·7	0·686	43·3	11·0
38 × 175	6·65	194	42·1	17·0	0·800	50·5	11·0
38 × 200	7·60	253	48·1	25·3	0·915	57·7	11·0
38 × 225	8·55	321	54·2	36·1	1·03	65·0	11·0
40 × 75	3·00	37·5	20·0	1·41	0·400	21·7	11·5
40 × 100	4·00	66·7	26·7	3·33	0·533	28·9	11·5
40 × 125	5·00	104	33·3	6·51	0·667	36·1	11·5
40 × 150	6·00	150	40·0	11·2	0·800	43·3	11·5
40 × 175	7·00	204	46·7	17·9	0·933	50·5	11·5
40 × 200	8·00	267	53·3	26·7	1·07	57·7	11·5
40 × 225	9·00	338	60·0	38·0	1·20	65·0	11·5

Table A.4 (*contd*)

Basic size	Area	Section modulus		Second moment of area		Radius of gyration	
		About x–x	About y–y	About x–x	About y–y	About x–x	About y–y
mm	10^3 mm²	10^3 mm³	10^3 mm³	10^6 mm⁴	10^6 mm⁴	mm	mm
44 × 75	3·30	41·2	24·2	1·55	0·532	21·7	12·7
44 × 100	4·40	72·3	32·3	3·67	0·710	28·9	12·7
44 × 125	5·50	115	40·3	7·16	0·887	36·1	12·7
44 × 150	6·60	165	48·4	12·4	1·06	43·3	12·7
44 × 175	7·70	225	56·5	19·7	1·24	50·5	12·7
44 × 200	8·80	293	64·5	29·3	1·42	57·7	12·7
44 × 225	9·90	371	72·6	41·8	1·60	65·0	12·7
44 × 250	11·0	458	80·7	57·3	1·77	72·2	12·7
44 × 300	13·2	660	96·8	99·0	2·13	86·6	12·7
50 × 75	3·75	46·9	31·2	1·76	0·781	21·7	14·4
50 × 100	5·00	83·3	41·7	4·17	1·04	28·9	14·4
50 × 125	6·25	130	52·1	8·14	1·30	36·1	14·4
50 × 150	7·50	188	62·5	14·1	1·56	43·3	14·4
50 × 175	8·75	255	72·9	22·3	1·82	50·5	14·4
50 × 200	10·0	333	83·3	33·3	2·08	57·7	14·4
50 × 225	11·2	422	93·8	47·5	2·34	65·0	14·4
50 × 250	12·5	521	104	65·1	2·60	72·2	14·4
50 × 300	15·0	750	125	112	3·12	86·6	14·4
63 × 100	6·30	105	66·2	5·25	2·08	28·9	18·2
63 × 125	7·88	164	82·7	10·3	2·60	36·1	18·2
63 × 150	9·45	236	99·2	17·7	3·13	43·3	18·2
63 × 175	11·0	322	116	28·1	3·65	50·5	18·2
63 × 200	12·6	420	132	42·0	4·17	57·7	18·2
63 × 225	14·2	532	149	59·8	4·69	65·0	18·2
75 × 100	7·50	125	93·8	6·25	3·52	28·9	21·7
75 × 125	9·38	195	117	12·2	4·39	36·1	21·7
75 × 150	11·2	281	141	21·1	5·27	43·3	21·7
75 × 175	13·1	383	164	33·5	6·15	50·5	21·7
75 × 200	15·0	500	188	50·0	7·03	57·7	21·7
75 × 225	16·9	633	211	71·2	7·91	65·0	21·7
75 × 250	18·8	781	234	97·7	8·79	72·2	21·7
75 × 300	22·5	1120	281	169	10·5	86·6	21·7
100 × 100	10·0	167	167	8·33	8·33	28·9	28·9
100 × 150	15·0	375	250	28·1	12·5	43·3	28·9
100 × 200	20·0	667	333	66·7	16·7	57·7	28·9
100 × 250	25·0	1040	417	130	20·8	72·2	28·9
100 × 300	30·0	1500	500	225	25·0	86·6	28·9
150 × 150	22·5	562	562	42·2	42·2	43·3	43·3
150 × 200	30·0	1000	750	100	56·2	57·7	43·3
150 × 300	45·0	2250	1120	338	84·4	86·6	43·3
200 × 200	40·0	1330	1330	133	133	57·7	57·7
250 × 250	62·5	2600	2600	326	326	72·2	72·2
300 × 300	90·0	4500	4500	675	675	86·6	86·6

A.5 SAFE WORKING LOADS FOR PROPS TO BS 4074

Figure A.5[23] gives the safe working load for props when erected to the maximum tolerances permitted, see section A.6. When nominally concentric loading can be ensured, e.g. by the use of wedges in a forkhead, the higher values in (b) may be used.

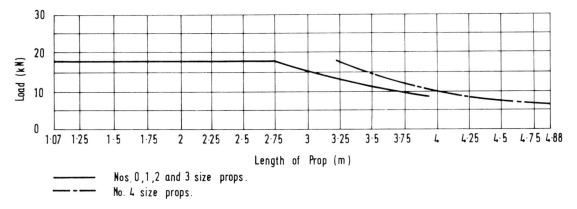

(a) Safe working loads for props erected 1.5° maximum out of plumb and with up to 25 mm maximum eccentricity of loading.

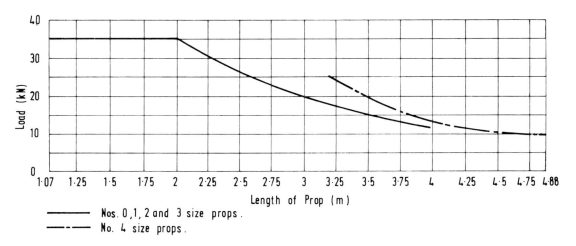

(b) Safe working loads for props erected 1.5° maximum out of plumb with concentric loading ensured.

Figure A.5 Safe working loads for props

A.6 ERECTION TOLERANCES AND WORKMANSHIP

BS 5975 recommends the following criteria:

Adjustable steel props and forkheads
(a) Props should be undamaged and not visibly bent.
(b) Props should be plumb within 1.5° of vertical (i.e. not exceeding 25 mm out-of-vertical over a height of 1 m).
(c) Props should be placed centrally under the member to be supported and over any member supporting the prop, with no eccentricity in excess of 25 mm.

(d) Forkheads where used should be rotated to centralise the bearer they support. Where beams terminate in a forkhead they should extend past the centre joint of the forkhead by at least 50 mm. Alternatively, where timbers butt in a forkhead, the joint should be within 15 mm of the centre of the forkhead.

Tube and coupler falsework

(a) The tubes used in falsework should be undamaged, not visibly bent or creased and have smooth square cut ends. Other components should also be undamaged.
(b) Verticals should be plumb within 15 mm over 2 m of height, subject to a maximum displacement from the vertical of 25 mm.
(c) Vertical members should be placed centrally under the members to be supported and over the member supporting them with no eccentricity exceeding 25 mm.
(d) Adjustable forkheads and baseplates should be adequately laced or braced where their extension exceeds 300 mm, unless an alternative figure is specified. The bracing tubes should be attached close to the fork or baseplate and to an adjacent vertical member, close to the lacing.
(e) Tubes should have end-to-end joints in adjacent tubes staggered. Sleeve couplers should be used in preference to joint pins for axial connections.
(f) The centre lines of tubes at a node point should be as close together as possible, and never more than 150 mm apart.
(g) Sole plates used to distribute loads on to foundation soils should normally be set horizontally within a tolerance not exceeding 25 mm in a length of 1 m.
(h) Top arrangements as (d) previously.

Appendix B
Approximate Analysis of Trusses by the Method of Sections

To obtain an accurate analysis of complex and multi-redundant trusses a finite element analysis is generally required. Often in falsework structures such sophistication is unnecessary and a reliable and conservative approximate analysis will suffice. One simple and versatile method of analysing statically determinate frames is the method of sections.

In order to use the method of sections in an approximate analysis of frames with redundant bracing members a commonly employed simplification is to remove compression diagonals from the analysis where cross diagonals are employed. The reasoning behind this is twofold. Firstly, it is assumed that these compression elements flex slightly, thus shedding their forces to the tension diagonals and secondly, this modification renders a statically determinate frame for analysis purposes. For example the frame in fig. B.1a is reduced to that of Fig. B.1b. To determine reaction forces, consider moments about one support, thus eliminating that reaction from the immediate calculations, and evaluate say the reaction at 'A'. Now sum vertical forces to

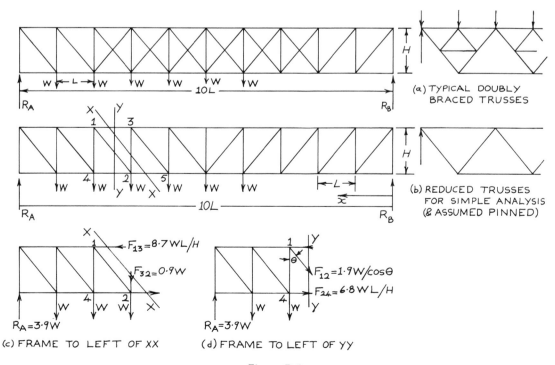

(a) TYPICAL DOUBLY BRACED TRUSSES

(b) REDUCED TRUSSES FOR SIMPLE ANALYSIS (& ASSUMED PINNED)

(c) FRAME TO LEFT OF XX

(d) FRAME TO LEFT OF YY

Figure B.1

find the reaction force at 'B'. The forces in the booms or bracing can now be obtained by the method of sections as follows:

Force in top boom member $1-3$–Cut the frame on XX and take moments about '2' for the left-hand side of the frame (fig. B.1c).

Force in bottom boom member $4-2$–Cut the frame on YY and take moments about '1' for the left-hand side of the frame (fig. B.1d).

Force in diagonal $1-2$–Cut the frame on YY and sum the vertical force components for the left-hand side of the frame (fig. B.1d).

Force in vertical $2-3$–Cut the frame on XX and sum the vertical forces for the left-hand side of the frame (fig. B.1c).

Note – If the sign of the derived force is negative this indicates that the chosen direction of the force was wrong. Simply alter the direction of the force arrow. The force magnitude remains unaltered.

The diagonals of a crossed-braced frame can be in tension or compression depending upon the loading pattern. Initial analyses can be carried out to determine the tensile diagonals for a given loading if required.

From fig. B.1b the reaction force at 'A' is

$$R_A = \frac{\Sigma Wx}{10L} = 3.9W$$

Therefore

$$R_B = \Sigma W - R_A = 2.1W$$

By moments about '2' the force in $1-3$ is (fig. B.1c):

$$F_{13} = \frac{8.7WL}{H} \quad \text{(compression)}$$

Force in $3-2$ is:

$$F_{32} = 3.9W - 3W = 0.9W$$

Force in $1-4$ is:

$$F_{14} = 1.9W = F_{12}\cos\theta$$

From fig. B.1d the force in $4-2$ by moments about '1' is:

$$F_{24} = \frac{6.8WL}{H} \quad \text{(tension)}$$

By summation of vertical forces the force in $1-2$ is:

$$F_{12} = \frac{1.9W}{\cos\theta} \quad \text{(tension)}$$

Force in $3-5$ is:

$$F_{35} = \frac{0.9W}{\cos\theta} = \frac{F_{32}}{\cos\theta}$$

Appendix C
Approximate Frame Deflections by the Flexibility Method

A conservative estimate of the deflections of a frame or lattice girder can be obtained by flexibility methods. As a first step the frame is simplified if necessary by the same method as indicated in appendix B and the axial force in each member obtained for the loading under consideration. The axial force in each member is now found for the case of all acting loads removed from the frame and a unit point load applied to the frame at the point and in the direction in which it is desired to estimate the frame deflection under the applied loads. The value of the deflection at the point and for the unit load direction is now obtained from:

$$\delta = (\sum_{1}^{n} P_{i1} P_{i2}) L_i / A_i E_i$$

where P_{i1} = force in member i from loads applied to the lattice girder
P_{i2} = force in member i from the applied unit load
L_i = length of member i
A_i = cross-sectional area of member i
E_i = modulus of elasticity of member i
n = number of members in the lattice.

Figure C.1a shows a loaded lattice girder and the force in each member for the given loading. If the vertical deflection has to be evaluated at point '6' then a unit load is applied vertically at point '6' and the forces in the members evaluated as given in fig. C.1b. If all members are of area 'A', length 'L' and modulus of elasticity 'E' then for $L = 3$ m, $A = 23.4$ cm^2, $E = 210$ kN/mm^2, and $W = 30$ kN

$$\delta_6 = \frac{(\sum_{1}^{n} P_{i1} P_{i2})L}{AE} = \frac{6.405LW}{AE} = 1.17 \text{ mm}$$

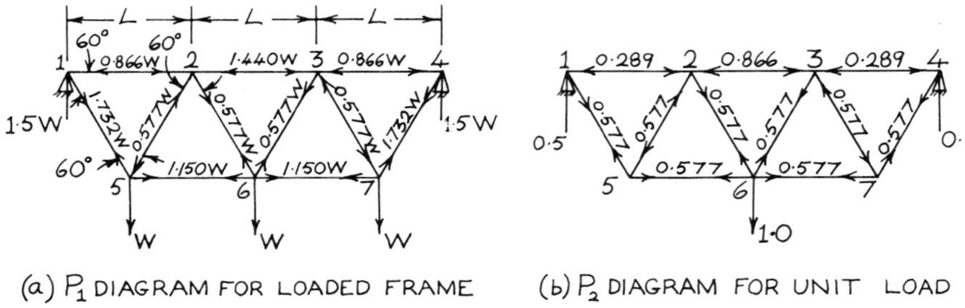

(a) P_1 DIAGRAM FOR LOADED FRAME   (b) P_2 DIAGRAM FOR UNIT LOAD

Figure C.1

Appendix D
Simple Analysis of Members for Torsion

Wherever possible torsion of members is to be avoided in falsework structures by careful design and erection and loading control. Where torsion is unavoidable analysis can often be carried out by the following simple methods.

(1) Simple open sections with no warping restraint
(a) For a long member where the breadth of any rectangular component (fig. D.1) is much greater than the thickness, then the twist per unit length is given by:

$$T/\Theta_1 = C = \Sigma Gbt^3/3$$

where I = applied torque
$\quad\quad \Theta_1$ = twist per unit length
$\quad\quad\; C$ = torsional stiffness
$\quad\quad\; G$ = shear modulus,

and the maximum shear stress from torsion is given by:

$$f_{qT_{max}} = G\Theta_1 t_{max}$$

At a re-entrant corner the stresses from torsion are magnified by the concentration of shear flow and the stress here is given by:

$$f_{rT} = 1.74 f_{qT_{max}} \sqrt[3]{(t_{max}/\text{radius of corner fillet})}$$

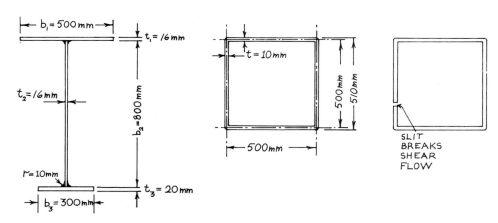

Figure D.1

(b) If the member is such that the breadth of a rectangle is not great in comparison to the thickness then the above may be modified using, $T/\Theta_1 = C = Gbt^3/k$ where k assumes a value dependent upon b/t:

b/t	∞	10.0	5.0	3.0	2.5	2.0	1.5	1.0
k	3.000	3.205	3.436	3.802	4.016	4.367	5.102	7.092

To illustrate the use of this analysis consider the plate girder shown in fig. D.1 which is subject to torsional forces. For $G = 80\,kN/mm^2$

$$\Sigma Gbt^3/3 = 80(500 \times 16^3 + 800 \times 16^3 + 300 \times 20^3)/3 = 2.06 \times 10^8\,kN/mm$$

The torsion from eccentric loading amounts to 5 kNm and the length of the cantilever is 2 m therefore the total twist is:

$$\Theta = 5 \times 10^3 \times 2 \times 10^3/2.06 \times 10^8 = 4.85 \times 10^{-2}\ radians$$

The maximum shear stress in a plate from torsion is:

$$f_{qT_{max}} = G\Theta_1 t_{max} = 80 \times 4.85 \times 10^{-2} \times 20/2 = 38.8\,N/mm^2$$

If the fillet radius of the welds is 10 mm then the corner shear stress is:

$$f_{rT} = 1.74 f_{qT_{max}} \sqrt[3]{(t_{max}/\text{fillet radius})} = 1.74 \times 38.8 \times \sqrt[3]{20/10} = 85\,N/mm^2$$

This illustrates the importance of avoiding torsion of such beams not only from a buckling and stability viewpoint but also because the stresses can be significant. In this case care has also to be taken during placement to ensure that the girder with unequal flanges is positioned the correct way up.

(2) Thin-walled hollow sections with no slits

$$T/\Theta_1 = C = 4GA^2t/L_p$$

where A = the area enclosed by the centre line of the section wall
  L_p = length of the centre line of the wall cross section

and
$$f_{qT} = T/2At$$

For tubes not of constant wall thickness, for thick walled or multi-cell tubes the analysis is more complex.

The box section shown in fig. D.1 is 2 m long and is subject to end torsion of 5 kNm. Note – the use of a square cross section avoids the possibility of incorrect placement as can occur with rectangular boxes.

$$T/\Theta_1 = 4GA^2t/L_p = 4 \times 80 \times (500 \times 500)^2 \times 10/(500 \times 4) = 1 \times 10^{11}\,kNmm^2$$

The total twist is given by:

$$\Theta = 5 \times 10^3/1 \times 10^{11} = 5 \times 10^{-8}\ radians$$

and the shear stress from torsion is:

$$f_{qT} = T/(2At) = 5 \times 10^6/(2 \times 500^2 \times 10) = 1\,N/mm^2$$

This illustrates the superior torsional stiffness of enclosed sections in torsion in comparison to open sections. If the tube were to be slit such that it no longer remained a completely closed section (fig. D.1) then the analysis and performance in torsion would be as for an open section.

Appendix E
Wind Force Data

Table E.1 Wind data

(i) Topography Factor S_1

Topography	Value of S_1
(a) All cases except those in (b) and (c) below	1.0
(b) Very exposed hill slopes and crests where acceleration of the wind is known to occur. Valleys shaped to produce a funnelling of the wind	1.1
(c) Steep-sided, enclosed valleys sheltered from all winds	0.9

(ii) Shielding factor

Spacing ratio	Aerodynamic solidity ratio*							
	0.1	0.2	0.3	0.4	0.5	0.6	0.7	0.8 & over
Up to 1.0	1.0	0.96	0.90	0.80	0.68	0.54	0.44	0.37
2.0	1.0	0.97	0.91	0.82	0.71	0.58	0.49	0.43
3.0	1.0	0.97	0.92	0.84	0.74	0.63	0.54	0.48
4.0	1.0	0.98	0.93	0.86	0.77	0.67	0.59	0.54
5.0	1.0	0.98	0.94	0.88	0.80	0.71	0.64	0.60
6.0 and over	1.0	0.99	0.95	0.90	0.83	0.75	0.69	0.66

*Aerodynamic solidity ratio = solidity ratio (ϕ) × a constant
where the constant = 1.6 for flat members
 = 1.2 for circular sections and for flat members in conjunction with such circular sections.

(iii) Values of q in N/m^2

V_s (m/s)	0	1.0	2.0	3.0	4.0	5.0	6.0	7.0	8.0	9.0
10	61	74	88	104	120	138	157	177	199	221
20	245	270	297	324	353	383	414	447	481	516
30	552	589	628	668	709	751	794	839	885	932
40	981	1030	1080	1130	1190	1240	1300	1350	1410	1470
50	1530	1590	1660	1720	1790	1850	1920	1990	2060	2130
60	2210	2280	2360	2430	2510	2590	2670	2750	2830	2920
70	3000									

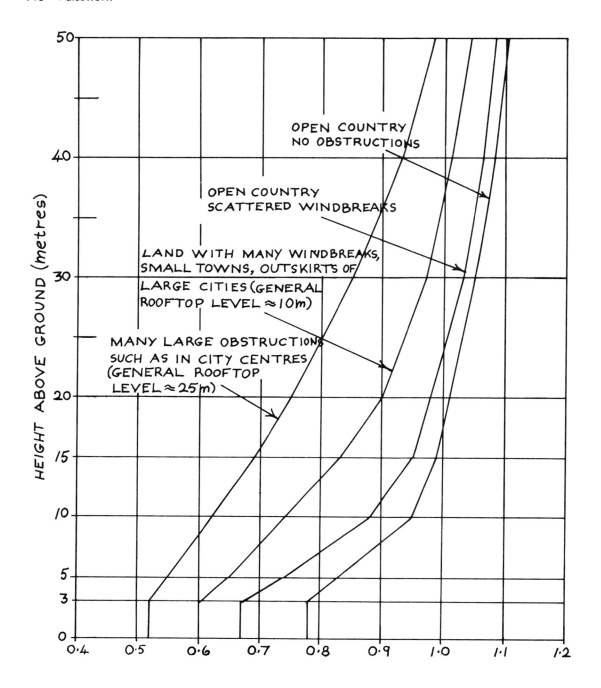

GROUND SURFACE FACTOR S₂ FOR WIND LOADING (INDIVIDUAL MEMBERS, CLASS A , CP3:CHAPTER V:PART 2)

Figure E.1

Table E.2 Wind force coefficients

The following values should be used in the general case. These assume a solidarity ratio of approximately 0.125 for frames and towers. Where the actual solidarity ratio varies by more than 0.10 from this value, reference should be made to CP3, Chapter V: Part 2: Wind Loading.

(a) Individual members:

Flat-sided members	$C_f = 2.0$
Circular sections	$C_f = 1.2$

(b) Single frames:

Flat-sided members	$C_f = 1.8$
Circular sections	$C_f = 1.2$

(c) Square lattice towers (any shielding effect allowed for):

Flat-sided members	$C_f = 3.7$
Circular sections	$C_f = 2.2$

(d) Triangular lattice towers (any shielding effect allowed for)

Flat-sided members	$C_f = 3.0$
Circular sections	$C_f = 1.7$

(e) Universal Beams and Columns $C_f = 1.6$

Table E.3 Values of wind speed factor, S_3 (for probability level 0.63)

Life of falsework in years	less than 2	2 to 5	5 to 10	over 10
Wind speed factor, S_3	0.77	0.83	0.88	1.00

Appendix F
General Design Aids and Data Tables

Table F.1 General falsework loadings

Routine working areas	$1.50 \, \text{kN/m}^2$
Access	$0.75 \, \text{kN/m}^2$
Storage areas	Individually assessed
Public access	$4.00 \, \text{kN/m}^2$
Pedestrian barriers	$0.75 \, \text{kN/m}$ in any direction
Vehicle crash barriers	$7.50 \, \text{kN/m}$
Vehicle loads	Axle or wheel load with impact allowance of 11.25 t maximum
Vertical dynamic loads	Motorised hoist – $1.25 \times$ static load Manual hoist – $1.10 \times$ static load
Horizontal dynamic loads	for $V < 2$ m/s – $0.10 \times$ static load for $V > 2$ m/s – $0.33 \times$ static load
Heaping of concrete on a 1 m² maximum area	Slab depth + twice slab depth heaping $= 3 \times d$ total
Concrete pumping force on falsework	$0.25 \times$ pumping pressure (N/mm²) $\times$ pipeline cross-sectional area (mm²)
Probable maximum pump pressures	Mechanical pumps – $5.00 \, \text{N/mm}^2$ Pneumatic placers – $0.70 \, \text{N/mm}^2$
Powder snow density	$80 \, \text{kg/m}^3$
Probable maximum snow loading	$1.5 \, \text{kN/m}^2$ in UK in drifting
Maximum ice density	$920 \, \text{kg/m}^3$
Probable maximum thickness of ice around members	25 mm

Table F.2 Friction coefficients (from table 18, BS 5975)

Lower load-accepting member	Upper load bearing member				
	Plain steel	Painted steel	Concrete	Softwood timber	Hardwood
Plain steel	0.15	0.10	0.10	0.20	0.10
Painted or oiled steel	0.10	0.00	0.00	0.20	0.00
Concrete	0.10	0.00	0.40	0.40	0.30
Softwood timber	0.20	0.20	0.40	0.40	0.30
Granular soil	0.30	0.30	0.40	0.30	0.30
Hardwood	0.10	0.00	0.30	0.30	0.30
Rubber	0.60				
Urethane	0.40				

Table F.3 Safe bearing capacities for soils (based on a metre wide footing at 0.6 m depth, for goundwater $< B$ below base half values)

Soil type	Safe bearing capacity (kN/m^2)
Compact well-graded sand or gravel sand	>600
Loose sand/gravel	<200
Medium dense gravel	200 to 600
Compact sand	>300
Medium dense sand	100 to 300
Loose sand	<100
Shaly/very stiff boulder clay	300 to 600
Stiff clay	150 to 300
Firm clay	75 to 150
Very soft clays/silts/peat and organic soils	No recognised capacity
Made up ground/fill material	Control required if used for support
Rock (depending upon type, fissures, caving, voids, lubrication)	Under 600 to 10 000

Table F.4 Thermal expansion coefficients

Material	Expansion coefficient ($\times 10^6$)	
	per °C	per °F
Structural steel	11.7	6.5
Average for concrete	9.9	5.5
Quartz aggregate	11.9	6.6
Sandstone aggregate	11.7	6.5
Gravel	10.8 to 12	$\simeq 6.0$
Granite	9.5	5.3
Basalt	8.6	4.8
Limestone	6.8	3.8
Mild steel	10.0	5.6

Table F.5 Steel properties

Poisson's ratio	0.28
Shear modulus	80 kN/mm^2
Modulus of elasticity	210 kN/mm^2
Density	7850 kg/m^3

Table F.6 Mass of concrete

Type	Mass (kg/m^3)
Average	2300
Heavy aggregate	3200
Lightweight aggregate	1150 to 1600
Average with 3% steel	2550
Average with 4% steel	2610
Average with 5% steel	2660
Average with 6% steel	2720

Table F.7 Metric bolt data in range M10 to M36 for course threaded bolts

Size	Thread pitch (mm)	Shank cross-sectional area (mm^2)		Width across corners (mm)	Width across flats (mm)	Clearance hole size (mm)	
		gross	root of thread			fine	medium
M10	1.50	79	58	19.6	17.0	10.5	11.0
M12	1.75	113	83	21.9	19.0	13.0	14.0
M16	2.00	201	157	27.7	24.0	17.0	18.0
M18*	2.50	254	198	31.2	27.0	19.0	20.0
M20	2.50	314	245	34.6	30.0	21.0	22.0
M22*	2.50	380	303	36.9	32.0	23.0	24.0
M24	3.00	452	353	41.6	36.0	25.0	26.0
M27*	3.00	573	459	47.3	41.0	28.0	30.0
M30	3.50	707	561	53.1	46.0	31.0	33.0
M33*	3.50	855	694	57.7	50.0	34.0	36.0
M36	4.00	1018	817	63.5	55.0	37.0	39.0

* Non-preferred sizes.

Table F.8 Stress and design data for steel

(a) Edge distance of holes

Diameter (mm)	Distance to sheared or hand flame cut edge (mm)	Distance to rolled, machine cut sawn or planed edge (mm)
39	68	62
36	62	56
33	56	50
30	50	44

Table F.8 (*contd*)

(b) Allowable bearing stress, p_b (N/mm^2) for grade 43 steel

Plates, sections and bars	190
Hot rolled hollow sections	190

(c) Allowable maximum shear stress, p_q (N/mm^2) for grade 43 steel

Plates, bars and sections	115
over 40 mm thick	105
'I' beams and plate girders with unstiffened webs	100

(d) Allowable stress in axial tension, p_t (N/mm^2) for grade 43 steel

Rolled 'I' beams and channels	155
Universal beams and columns	
not greater than 40 mm thick	155
over 40 mm thick	140
Other sections, plates and bars	
not greater than 40 mm thick	155
over 40 mm thick	140

Table F.9 Allowable stress p_c on gross section for axial compression in grade 43 steel sections

l/r	\multicolumn{10}{c}{p_c (N/mm^2) for grade 43 steel}

l/r	0	1	2	3	4	5	6	7	8	9
0	155	155	154	154	153	153	153	152	152	151
10	151	151	150	150	149	149	148	148	148	147
20	147	146	146	146	145	145	144	144	144	143
30	143	142	142	142	141	141	141	140	140	139
40	139	138	138	137	137	136	136	136	135	134
50	133	133	132	131	130	130	129	128	127	126
60	126	125	124	123	122	121	120	119	118	117
70	115	114	113	112	111	110	108	107	106	105
80	104	102	101	100	99	97	96	95	94	92
90	91	90	89	87	86	85	84	83	81	80
100	79	78	77	76	75	74	73	72	71	70
110	69	68	67	66	65	64	63	62	61	61
120	60	59	58	57	56	56	55	54	53	53
130	52	51	51	50	49	49	48	48	47	46
140	46	45	45	44	43	43	42	42	41	41
150	40	40	39	39	38	38	38	37	37	36
160	36	35	35	35	34	34	33	33	33	32
170	32	32	31	31	31	30	30	30	29	29
180	29	28	28	28	28	27	27	27	26	26
190	26	26	25	25	25	25	24	24	24	24
200	24	23	23	23	23	22	22	22	22	22
210	21	21	21	21	21	20	20	20	20	20
220	20	19	19	19	19	19	19	18	18	18
230	18	18	18	18	17	17	17	17	17	17
240	17	16	16	16	16	16	16	16	16	15

Intermediate values may be obtained by linear interpolation.

Note – For material over 40 mm thick, other than rolled I-beams or channels, and for Universal columns of thicknesses exceeding 40 mm, the limiting stress is 140 N/mm^2.

Sections with $L/r > 180$ to be used with caution.

Table F.10 Allowable stress p_{bc} in bending (N/mm^2) for beams of grade 43 steel

l/r_y	D/T							
	10	15	20	25	30	35	40	50
90	165	165	165	165	165	165	165	165
95	165	165	165	163	163	163	163	163
100	165	165	165	157	157	157	157	157
105	165	165	160	152	152	152	152	152
110	165	165	156	147	147	147	147	147
115	165	165	152	141	141	141	141	141
120	165	162	148	136	136	136	136	136
130	165	155	139	126	126	126	126	126
140	165	149	130	115	115	115	115	115
150	165	143	122	104	104	104	104	104
160	163	136	113	95	94	94	94	94
170	159	130	104	91	85	82	82	82
180	155	124	96	87	80	76	72	71
190	151	118	93	83	77	72	68	62
200	147	111	89	80	73	68	64	59
210	143	105	87	77	70	65	61	55
220	139	99	84	74	67	62	58	52
230	134	95	81	71	64	59	55	49
240	130	92	78	69	61	56	52	47
250	126	90	76	66	59	54	50	44
260	122	88	74	64	57	52	48	42
270	118	86	72	62	55	50	46	40
280	114	84	70	60	53	48	44	39
290	110	82	68	58	51	46	42	37
300	106	80	66	56	49	44	41	36

Intermediate values may be obtained by linear interpolation.

NOTE. For materials over 40 mm thick the stress shall not exceed 150 N/mm^2.

Table F.11 Scaffold tube data

Wall thickness	4.00 mm (+0.8/−0.4 mm tolerance)
Outside diameter	48.3 mm (±0.5 mm tolerance)
Nominal mass	4.37 kg/m (−7.5% maximum)
Cross-sectional area	557 mm^2
Second moment of area about a diameter	138 000 mm^4
Section modulus	5700 mm^3
Ultimate tensile strength	340 N/mm^2
Yield stress	210 N/mm^2
Allowable compressive stress	124 N/mm^2 (maximum)
Radius of gyration	15.7 mm
Modulus of elasticity	210 kN/mm^2
Permissible shear stress	93 N/mm^2
Straightness	Maximum deviation 1/1600 × length

Table F.12 Bolt stress data for bolts of designation 4.6 (N/mm²)

Axial tensile stress on net area of bolts and tension rods	130
Shear stress on the gross area of bolts	
close tolerance turned bolts	95
black bolts	80
In bearing (double shear)	
close tolerance turned bolts	300
black bolts	200

Table F.13 Maximum permissible axial stresses and loads in steel scaffold tubes

Effective length, l (mm)	Slenderness ratio, l/r	New tubes		Used tubes	
		Permissible axial compressive stress, p_c (N/mm²)	Permissible axial load (kN)	Permissible axial compressive stress, p_c (N/mm²)	Permissible axial load (kN)
0		127	70.7	108	60.1
250	15.9	123	68.5	105	58.2
500	31.8	119	66.2	101	56.3
750	47.8	113	63.0	96.2	53.6
1000	63.7	104	57.7	88.1	49.1
1250	79.6	90.3	50.3	76.8	42.8
1500	95.5	75.4	42.0	64.1	35.7
1750	111.5	61.4	34.2	52.2	29.1
2000	127.4	50.0	27.9	42.5	23.7
2250	143.3	40.9	22.8	34.8	19.4
2500	159.2	34.0	18.9	28.9	16.1
2750	175.2	28.7	16.0	24.4	13.6
3000	191.1	24.2	13.5	20.6	11.5
3250	207.0	20.9	11.6	17.8	9.9
3500	222.9	18.1	10.1	15.4	8.6
3750	238.8	15.9	8.8	13.5	7.5
4000	254.8	14.1	7.9	12.0	6.7

Note – Values of l/r greater than 180 should be used with great caution.
   – The maximum effective length for used tubes is 1900 mm.

Table F.14 Masses of corrugated steel sheeting

Standard thickness (mm)	Approximate equivalent standard wire gauge	Number of corrugations (nominal cover width is given in brackets*)									
		8/3 (610 mm)		10/3 (762 mm)		10 1/2/3 (800 mm)		12/3 (914 mm)		12 1/2/3 (952 mm)	
		(kg/m)	(m/tonne)	(kg/m)	(m/tonne)	(kg/m)	(m/tonne)	(kg/m)	(m/tonne)	(kg/m)	(m/tonne)
0.425		2.50	400	3.05	328	3.18	315	3.59	278	3.80	263
0.50	26	2.94	341	3.59	279	3.74	267	4.23	236	4.47	224
0.60	24	3.52	284	4.31	232	4.49	223	5.07	197	5.36	187
0.70	22	4.11	243	5.02	199	5.23	191	5.92	169	6.25	160
0.80		4.70	213	5.74	174	5.98	167	6.77	148	7.15	140
0.90		5.28	192	6.45	156	6.73	151	7.61	133	8.04	126
1.00	20	5.87	170	7.17	139	7.48	134	8.46	118	8.93	112
1.20	18	7.05	142	8.61	116	8.97	111	10.15	99	10.72	93
1.60		9.40	106	11.48	87	11.96	84	13.53	74	14.29	70
2.00		11.75	85	14.35	70	14.95	67	16.92	59	17.87	56

*The nominal cover width is the width of the sheeting after corrugation, and is subject to manufacturing tolerances.

Table F.15 Safe loadings for single couplers and components

Coupler or component	Loading	Safe load (kN)
Swivel coupler	Slip along tube	6.250
90° coupler	Slip along tube	6.250
Putlog coupler	Axial pull out force	0.625
Sleeve couplers	Tension	3.100
Joint pins in expanding spigot couplers	Shear capacity	21.000
Putlog ends	Shear	1.120
Adjustable base plate	Axial compression	30.000

Table F.16 Effective lengths of compression members

Deformation form	Form of restraint	Effective length, l
	Effectively held in position and direction at both ends	$0.7L$*
	Effectively held in position at both ends and restrained in direction at one end	$0.85L$*
	Effectively held in position at both ends but not restrained in direction	$1.0L$
	Effectively held in position and restrained in direction at one end and partially restrained in direction but not held in position at the other end	$1.5L$
	Effectively held in position and restrained in direction at one end but not held in position or restrained at the other end	$2.0L$

*These values are not normally appropriate in tube and coupler systems.

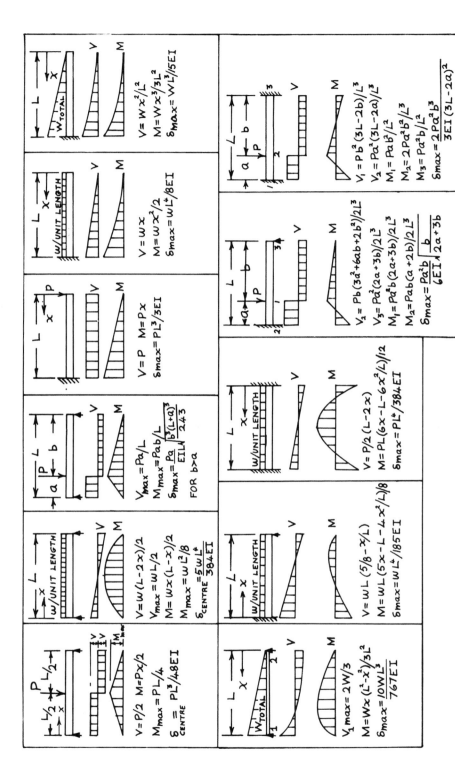

Figure F.1 Data for prismatic beams

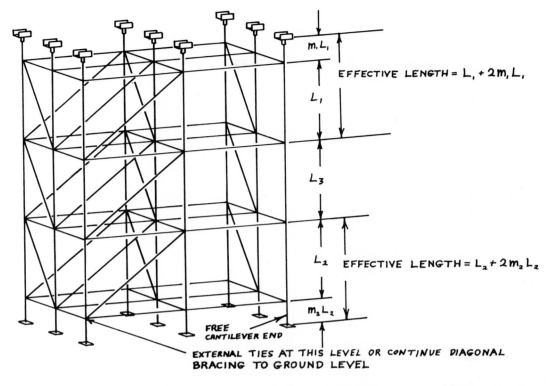

Figure F.2 Diagrammatic configuration showing effective lengths in tube and coupler scaffolding

Table F.17 Effective length of cantilever beams without intermediate lateral restraint

Restraint conditions		Effective length, l	
at support	at tip	normal loading condition	destabilising loading condition*
Built in, i.e. no lateral or rotational displacement	Free	0.8L	1.4L
	Top flange held against lateral displacement	0.7L	1.4L
	Both flanges held against lateral displacement	0.6L	0.6L
Continuous, and both flanges held against lateral displacement but free to rotate in plan	Free	1.0L	2.5L
	Top flange held against lateral displacement	0.9L	2.5L
	Both flanges held against lateral displacement	0.8L	1.5L
Continuous, with top flange only held against lateral displacement	Free	3.0L	7.5L
	Top flange held against lateral displacement	2.7L	7.5L
	Both flanges held against lateral displacement	2.4L	4.5L

*The destabilising loading condition exists when the load is applied on the top flange, and both the load and the flange are free to move laterally.
Note – L is the projecting length of the cantilever.

Table F.18 Effective lengths of beams without intermediate lateral restraints

End restraint condition	Effective length
Compression flange fully restrained against rotation in plan at the supports	$0.85L$
Compression flange partially restrained against rotation in plan at supports	$1.05L$
Compression flange unrestrained rotation in plan at the supports	$1.20L$
Bottom flange restrained at support against longitudinal twist but not otherwise restrained elsewhere in the span	$1.2(L + 2D)$

Note – If the load and top flange are restrained laterally then the above values may be multiplied by 0.8.

Table F.19 Effective lengths of load-bearing stiffeners

Restraint conditions	Effective length
Loaded flange held against lateral deflection relative to the other flange and against rotation in the plane of the stiffener	$0.7L_{stiffener}$
Loaded flange held against lateral deflection relative to other flange but not against rotation	$1.0L_{stiffener}$
Loaded flange held against rotation in the plane of the stiffener but not against lateral movement relative to the other flange	$1.2L_{stiffener}$
Loaded flange not restrained against rotation in the plane of the stiffener or against lateral deflection relative to the other flange.	$2.0L_{stiffener}$

Table F.20 Effective lengths and slenderness ratios of unstiffened webs in compression

Deformation form	Form of restraint	Effective length, l	Ratio l/r
	Restraint of web ends against both relative movement and rotation	$\simeq 0.7D$	$2.4D/t_w$
	Restraint of web ends against relative lateral movement but no rotation restraint	$\simeq 1.0D^*$	$3.5D/t_w$
	Restraint of web ends against rotation but not against relative lateral movement	$\simeq 1.2D$	$4.2D/t_w$
	Rotation and relative lateral restraint of one web but no restraint of the other web	$\simeq 2.0D^*$	$7.0D/t_w$

*If the ends are unrestrained against rotation the slenderness ratio should be based on the distance between the effective centres of rotation. This may necessitate the use of effective lengths greater than D or $2.0D$.

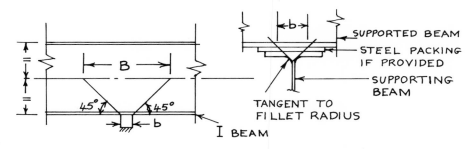

b = STIFF BEARING LENGTH (UPWARDS OR DOWNWARDS ACTIONS)
B = WEB WIDTH TO RESIST BUCKLING AT LOAD OR BEARING POINTS (SPREAD REDUCED WHEN 45° LINE MEETS A BEAM END)

Figure F.3 Web width to resist buckling at load or bearing positions

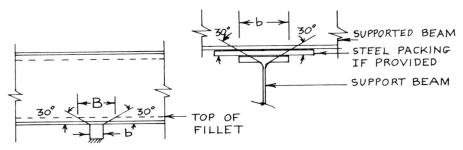

b = STIFF BEARING LENGTH

B = WEB WIDTH TO ACCEPT BEARING FORCES AT LOAD
TRANSFER POSITIONS (SPREAD REDUCED WHEN
30° LINE MEETS A BEAM END)

Figure F.4 Web width to accept bearing forces at load transfer positions

Table F.21 Square hollow sections

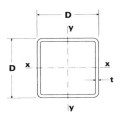

Designation		Mass per metre (kg)	Area of section (cm²)	Moment of inertia (cm⁴)	Radius of gyration (cm)	Elastic modulus (cm³)	Torsion constant (cm⁴)
Size $D \times D$ (mm)	Thickness t (mm)						
30 × 30	3.2	2.65	3.38	4.00	1.09	2.67	6.45
40 × 40	3.2	3.66	4.66	10.4	1.50	5.22	16.5
	4.0	4.46	5.68	12.1	1.46	6.07	19.5
50 × 50	3.2	4.66	5.94	21.6	1.91	8.62	33.8
	4.0	5.72	7.28	25.5	1.87	10.2	40.4
	5.0	6.97	8.88	29.6	1.83	11.9	47.6
60 × 60	3.2	5.67	7.22	38.7	2.31	12.9	60.1
	4.0	6.97	8.88	46.1	2.28	15.4	72.4
	5.0	8.54	10.9	54.4	2.24	18.1	86.3
70 × 70	3.6	7.46	9.50	69.5	2.70	19.9	108
	5.0	10.1	12.9	90.1	2.64	25.7	142
80 × 80	3.6	8.59	10.9	106	3.11	26.5	164
	5.0	11.7	14.9	139	3.05	34.7	217
	6.3	14.4	18.4	165	3.00	41.3	261
90 × 90	3.6	9.72	12.4	154	3.52	34.1	237
	5.0	13.3	16.9	202	3.46	45.0	315
	6.3	16.4	20.9	242	3.41	53.9	381

Table F.21 (*contd.*)

Size $D \times D$ (mm)	Thickness t (mm)	Mass per metre (kg)	Area of section (cm²)	Moment of inertia (cm⁴)	Radius of gyration (cm)	Elastic modulus (cm³)	Torsion constant (cm⁴)
100 × 100	4.0	12.0	15.3	234	3.91	46.8	361
	5.0	14.8	18.9	283	3.87	56.6	439
	6.3	18.4	23.4	341	3.81	68.2	533
	8.0	22.9	29.1	408	3.74	81.5	646
	10.0	27.9	35.5	474	3.65	94.9	761
120 × 120	5.0	18.0	22.9	503	4.69	83.8	775
	6.3	22.3	28.5	610	4.63	102	949
	8.0	27.9	35.5	738	4.56	123	1 159
	10.0	34.2	43.5	870	4.47	145	1 381
150 × 150	5.0	22.7	28.9	1 009	5.91	135	1 548
	6.3	28.3	36.0	1 236	5.86	165	1 907
	8.0	35.4	45.1	1 510	5.78	201	2 348
	10.0	43.6	55.5	1 803	5.70	240	2 829
	12.5	53.4	68.0	2 125	5.59	283	3 372
	16.0	66.4	84.5	2 500	5.44	333	4 029
180 × 180	6.3	34.2	43.6	2 186	7.08	243	3 357
	8.0	43.0	54.7	2 689	7.01	299	4 156
	10.0	53.0	67.5	3 237	6.92	360	5 041
	12.5	65.2	83.0	3 856	6.82	428	6 062
	16.0	81.4	104	4 607	6.66	512	7 339
200 × 200	6.3	38.2	48.6	3 033	7.90	303	4 647
	8.0	48.0	61.1	3 744	7.83	374	5 770
	10.0	59.3	75.7	4 525	7.74	452	7 020
	12.5	73.0	93.0	5 419	7.63	542	8 479
	16.0	91.5	117	6 524	7.48	652	10 330
250 × 250	6.3	48.1	61.2	6 049	9.94	484	9 228
	8.0	60.5	77.1	7 510	9.87	601	11 511
	10.0	75.0	95.5	9 141	9.78	731	14 086
	12.5	92.6	118	11 050	9.68	884	17 139
	16.0	117	149	13 480	9.53	1 078	21 109
300 × 300	10.0	90.7	116	16 150	11.8	1 077	24 776
	12.5	112	143	19 630	11.7	1 309	30 290
	16.0	142	181	24 160	11.6	1 610	37 566
350 × 350	10.0	106	136	26 050	13.9	1 489	39 840
	12.5	132	168	31 810	13.8	1 817	48 869
	16.0	167	213	39 370	13.6	2 250	60 901
400 × 400	10.0	122	156	39 350	15.9	1 968	60 028
	12.5	152	193	48 190	15.8	2 409	73 815

Table F.22 Circular hollow sections

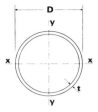

Designation							
Outside diameter D (mm)	Thickness t (mm)	Mass per metre (kg)	Area of section (cm²)	Moment of inertia (cm⁴)	Radius of gyration (cm)	Elastic modulus (cm³)	Torsional constants (cm⁴)
33.7	4.0	2.93	3.73	4.19	1.06	2.49	8.38
42.4	4.0	3.79	4.83	8.99	1.36	4.24	18.0
48.3	4.0	4.37	5.57	13.8	1.57	5.70	27.5
60.3	3.2	4.51	5.74	23.5	2.02	7.78	46.9
	4.0	5.55	7.07	28.2	2.00	9.34	56.3
	5.0	6.82	8.69	33.5	1.96	11.1	67.0
76.1	3.2	5.75	7.33	48.8	2.58	12.8	97.6
	4.0	7.11	9.06	59.1	2.55	15.5	118
	5.0	8.77	11.2	70.9	2.52	18.6	142
88.9	3.2	6.76	8.62	79.2	3.03	17.8	158
	4.0	8.38	10.7	96.3	3.00	21.7	193
	5.0	10.3	13.2	116	2.97	26.2	233
114.3	3.6	9.83	12.5	192	3.92	33.6	384
	5.0	13.5	17.2	257	3.87	45.0	514
	6.3	16.8	21.4	313	3.82	54.7	625
139.7	5.0	16.6	21.2	481	4.77	68.8	961
	6.3	20.7	26.4	589	4.72	84.3	1 177
	8.0	26.0	33.1	720	4.66	103	1 441
	10.0	32.0	40.7	862	4.60	123	1 724
168.3	5.0	20.1	25.7	856	5.78	102	1 712
	6.3	25.2	32.1	1 053	5.73	125	2 107
	8.0	31.6	40.3	1 297	5.67	154	2 595
	10.0	39.0	49.7	1 564	5.61	186	3 128
193.7	5.4	25.1	31.9	1 417	6.66	146	2 834
	6.3	29.1	37.1	1 630	6.63	168	3 260
	8.0	36.6	46.7	2 016	6.57	208	4 031
	10.0	45.3	57.7	2 442	6.50	252	4 883
	12.5	55.9	71.2	2 934	6.42	303	5 869
	16.0	70.1	89.3	3 554	6.31	367	7 109

Table F.22 (*contd.*)

Designation		Mass per metre (kg)	Area of section (cm²)	Moment of inertia (cm⁴)	Radius of gyration (cm)	Elastic modulus (cm³)	Torsional constants (cm⁴)
Outside diameter D (mm)	Thickness t (mm)						
219.1	6.3	33.1	42.1	2 386	7.53	218	4 772
	8.0	41.6	53.1	2 960	7.47	270	5 919
	10.0	51.6	65.7	3 598	7.40	328	7 197
	12.5	63.7	81.1	4 345	7.32	397	8 689
	16.0	80.1	102	5 297	7.20	483	10 590
	20.0	98.2	125	6 261	7.07	572	12 520
273	8.0	52.3	66.6	5 852	9.37	429	11 700
	10.0	64.9	82.6	7 154	9.31	524	14 310
	12.5	80.3	102	8 697	9.22	637	17 390
	16.0	101	129	10 710	9.10	784	21 410
	20.0	125	159	12 800	8.97	938	25 600
	25.0	153	195	15 130	8.81	1 108	30 250
323.9	8.0	62.3	79.4	9 910	11.2	612	19 820
	10.0	77.4	98.6	12 160	11.1	751	24 320
	12.5	96.0	122	14 850	11.0	917	29 690
	16.0	121	155	18 390	10.9	1 136	36 780
	20.0	150	191	22 140	10.8	1 367	44 280
	25.0	184	235	26 400	10.6	1 630	52 800
355.6	8.0	68.6	87.4	13 200	12.3	742	26 400
	10.0	85.2	109	16 220	12.2	912	32 450
	12.5	106	135	19 850	12.1	1 117	39 700
	16.0	134	171	24 660	12.0	1 387	49 330
	20.0	166	211	29 790	11.9	1 676	59 580
	25.0	204	260	35 680	11.7	2 007	71 350
406.4	10.0	97.8	125	24 480	14.0	1 205	48 950
	12.5	121	155	30 030	13.9	1 478	60 060
	16.0	154	196	37 450	13.8	1 843	74 900
	20.0	191	243	45 430	13.7	2 236	90 860
	25.0	235	300	54 700	13.5	2 692	109 400
	32.0	295	376	66 430	13.3	3 269	132 900
457	10.0	110	140	35 090	15.8	1 536	70 180
	12.5	137	175	43 140	15.7	1 888	86 290
	16.0	174	222	53 960	15.6	2 361	107 900
	20.0	216	275	65 680	15.5	2 874	131 400
	25.0	266	339	79 420	15.3	3 476	158 800
	32.0	335	427	97 010	15.1	4 246	194 000
	40.0	411	524	114 900	14.8	5 031	229 900

PROPERTIES OF STEEL SECTIONS

UNIVERSAL BEAMS

DIMENSIONS AND PROPERTIES

Serial Size (mm)	Mass per metre (kg)	Depth of Section D (mm)	Width of Section B (mm)	Thickness Web t (mm)	Thickness Flange T (mm)	Root Radius r (mm)	Depth between Fillets d (mm)	Area of Section (cm²)	Moment of Inertia x–x (cm⁴)	Moment of Inertia y–y (cm⁴)	Radius of Gyration x–x (cm)	Radius of Gyration y–y (cm)	Elastic Modulus x–x (cm³)	Elastic Modulus y–y (cm³)
457 × 152	82	465.1	153.5	10.7	18.9	10.2	404.4	104.4	36160	1094	18.6	3.24	1555	142.5
	74	461.3	152.7	9.9	17.0	10.2	404.4	94.9	32380	963	18.5	3.18	1404	126.1
	67	457.2	151.9	9.1	15.0	10.2	404.4	85.3	28522	829	18.3	3.12	1248	109.1
	60	454.7	152.9	8.0	13.3	10.2	407.7	75.9	25464	795	18.3	3.23	1120	104.0
	52	449.8	152.4	7.6	10.9	10.2	407.7	66.5	21345	645	17.9	3.11	949.0	84.61
406 × 178	74	412.8	179.7	9.7	16.0	10.2	357.4	94.9	27279	1448	17.0	3.91	1322	161.2
	67	409.4	178.8	8.8	14.3	10.2	357.4	85.4	24279	1269	16.9	3.85	1186	141.9
	60	406.4	177.8	7.8	12.8	10.2	357.4	76.1	21520	1109	16.8	3.82	1059	124.7
	54	402.6	177.6	7.6	10.9	10.2	357.4	68.3	18576	922	16.5	3.67	922.8	103.8
406 × 152	74	416.3	153.7	10.1	18.1	10.2	357.4	94.8	26938	1047	16.9	3.32	1294	136.2
	67	412.2	152.9	9.3	16.0	10.2	357.4	85.3	23798	908	16.7	3.26	1155	118.8
	60	407.9	152.2	8.6	13.9	10.2	357.4	75.8	20619	768	16.5	3.18	1011	100.9
406 × 140	46	402.3	142.4	6.9	11.2	10.2	357.4	58.9	15603	500	16.3	2.92	775.6	70.26
	39	397.3	141.8	6.3	8.6	10.2	357.4	49.3	12408	373	15.9	2.75	624.7	52.61
381 × 152	67	388.6	154.3	9.7	16.3	10.2	333.2	85.4	21276	947	15.8	3.33	1095	122.7
	60	384.8	153.4	8.7	14.4	10.2	333.2	75.9	18632	815	15.7	3.27	968.4	106.2
	52	381.0	152.4	7.8	12.4	10.2	333.2	66.4	16046	686	15.5	3.21	842.3	89.96
356 × 171	67	364.0	173.2	9.1	15.7	10.2	309.1	85.3	19483	1278	15.1	3.87	1071	147.6
	57	358.6	172.1	8.0	13.0	10.2	309.1	72.1	16038	1026	14.9	3.77	894.3	119.2
	51	355.6	171.5	7.3	11.5	10.2	309.1	64.5	14118	886	14.8	3.71	794.0	103.3
	45	352.0	171.0	6.9	9.7	10.2	309.1	56.9	12052	730	14.6	3.58	684.7	85.39
356 × 127	39	352.8	126.0	6.5	10.7	10.2	309.1	49.3	10054	333	14.3	2.60	570.0	52.87
	33	348.5	125.4	5.9	8.5	10.2	309.1	41.7	8167	257	14.0	2.48	468.7	40.99
305 × 165	54	310.9	166.8	7.7	13.7	8.9	262.6	68.3	11686	988	13.1	3.80	751.8	118.5
	46	307.1	165.7	6.7	11.8	8.9	262.6	58.8	9924	825	13.0	3.74	646.4	99.54
	40	303.8	165.1	6.1	10.2	8.9	262.6	51.4	8500	691	12.9	3.67	559.6	83.71
305 × 127	48	310.4	125.2	8.9	14.0	8.9	262.6	60.8	9485	438	12.5	2.68	611.1	69.94
	42	306.6	124.3	8.0	12.1	8.9	262.6	53.1	8124	367	12.4	2.63	530.0	58.99
	37	303.8	123.5	7.2	10.7	8.9	262.6	47.4	7143	316	12.3	2.58	470.3	51.11
305 × 102	33	312.7	102.4	6.6	10.8	7.6	275.3	41.8	6482	189	12.5	2.13	414.6	37.00
	28	308.9	101.9	6.1	8.9	7.6	275.3	36.3	5415	153	12.2	2.05	350.7	30.01
	25	304.8	101.6	5.8	6.8	7.6	275.3	31.4	4381	116	11.8	1.92	287.5	22.85
254 × 146	43	259.6	147.3	7.3	12.7	7.6	216.2	55.0	6546	633	10.9	3.39	504.3	85.97
	37	256.0	146.4	6.4	10.9	7.6	216.2	47.4	5544	528	10.8	3.34	433.1	72.11
	31	251.5	146.1	6.1	8.6	7.6	216.2	39.9	4427	406	10.5	3.19	352.1	55.53
254 × 102	28	260.4	102.1	6.4	10.0	7.6	224.5	36.2	4004	174	10.5	2.19	307.6	34.13
	25	257.0	101.9	6.1	8.4	7.6	224.5	32.1	3404	144	10.3	2.11	264.9	28.23
	22	254.0	101.6	5.8	6.8	7.6	224.5	28.4	2863	116	10.0	2.02	225.4	22.84
203 × 133	30	206.8	133.8	6.3	9.6	7.6	169.9	38.0	2880	354	8.71	3.05	278.5	52.85
	25	203.2	133.4	5.8	7.8	7.6	169.9	32.3	2348	280	8.53	2.94	231.1	41.92

Table F.23

Table F.24

UNIVERSAL BEAMS

DIMENSIONS AND PROPERTIES

Serial Size	Mass per metre	Depth of Section D	Width of Section B	Thickness Web t	Thickness Flange T	Root Radius r	Depth between Fillets d	Area of Section	Moment of Inertia Axis x–x	Moment of Inertia Axis y–y	Radius of Gyration Axis x–x	Radius of Gyration Axis y–y	Elastic Modulus Axis x–x	Elastic Modulus Axis y–y
mm	kg	mm	mm	mm	mm	mm	mm	cm²	cm⁴	cm⁴	cm	cm	cm³	cm³
914 × 419	388	920.5	420.5	21.5	36.6	24.1	791.5	493.9	717325	42481	38.1	9.27	15586	2021
	343	911.4	418.5	19.4	32.0	24.1	791.5	436.9	623866	36251	37.8	9.11	13691	1733
914 × 305	289	926.6	307.8	19.6	32.0	19.1	819.2	368.5	503781	14793	37.0	6.34	10874	961.3
	253	918.5	305.5	17.3	27.9	19.1	819.2	322.5	435796	12512	36.8	6.23	9490	819.2
	224	910.3	304.1	15.9	23.9	19.1	819.2	284.9	375111	10425	36.3	6.05	8241	685.6
	201	903.0	303.4	15.2	20.2	19.1	819.2	256.1	324715	8632	35.6	5.81	7192	569.1
838 × 292	226	850.9	293.8	16.1	26.8	17.8	756.4	288.4	339130	10661	34.3	6.08	7971	725.9
	194	840.7	292.4	14.7	21.7	17.8	756.4	246.9	278833	8384	33.6	5.83	6633	573.6
	176	834.9	291.6	14.0	18.8	17.8	756.4	223.8	245412	7111	33.1	5.64	5879	487.6
762 × 267	197	769.6	268.0	15.6	25.4	16.5	681.2	250.5	239464	7699	30.9	5.54	6223	574.6
	173	762.0	266.7	14.3	21.6	16.5	681.2	220.2	204747	6376	30.5	5.38	5374	478.1
	147	753.9	265.3	12.9	17.5	16.5	681.2	187.8	168535	5002	30.0	5.16	4471	377.1
686 × 254	170	692.9	255.8	14.5	23.7	15.2	610.6	216.3	169843	6225	28.0	5.36	4902	486.8
	152	687.6	254.5	13.2	21.0	15.2	610.6	193.6	150015	5391	27.8	5.28	4364	423.7
	140	683.5	253.7	12.4	19.0	15.2	610.6	178.4	135972	4789	27.6	5.18	3979	377.5
	125	677.9	253.0	11.7	16.2	15.2	610.6	159.4	117700	3992	27.2	5.00	3472	315.5
610 × 305	238	633.0	311.5	18.6	31.4	16.5	531.6	303.5	207252	14973	26.1	7.02	6549	961.3
	179	617.5	307.0	14.1	23.6	16.5	531.6	227.7	151312	10571	25.8	6.81	4901	688.6
	149	609.6	304.8	11.9	19.7	16.5	531.6	189.9	124341	8471	25.6	6.68	4079	555.9
610 × 229	140	617.0	230.1	13.1	22.1	12.7	543.1	178.2	111673	4253	25.0	4.88	3620	369.6
	125	611.9	229.0	11.9	19.6	12.7	543.1	159.4	98408	3676	24.8	4.80	3217	321.1
	113	607.3	228.2	11.2	17.3	12.7	543.1	144.3	87260	3184	24.6	4.70	2874	279.1
	101	602.2	227.6	10.6	14.8	12.7	543.1	129.0	75549	2658	24.2	4.54	2509	233.6
610 × 178	91	602.5	178.4	10.6	15.0	12.7	547.1	115.9	63970	1427	23.5	3.51	2124	160.0
	82	598.2	177.8	10.1	12.8	12.7	547.1	104.4	55779	1203	23.1	3.39	1865	135.3
533 × 330	212	545.1	333.6	16.7	27.8	16.5	450.1	269.6	141682	16064	22.9	7.72	5199	963.2
	189	539.5	331.7	14.9	25.0	16.5	450.1	241.2	125618	14093	22.8	7.64	4657	849.6
	167	533.4	330.2	13.4	22.0	16.5	450.1	212.7	109109	12057	22.6	7.53	4091	730.3
533 × 210	122	544.6	211.9	12.8	21.3	12.7	472.7	155.6	76078	3208	22.1	4.54	2794	302.8
	109	539.5	210.7	11.6	18.8	12.7	472.7	138.4	66610	2755	21.9	4.46	2469	261.5
	101	536.7	210.1	10.9	17.4	12.7	472.7	129.1	61530	2512	21.8	4.41	2293	239.2
	92	533.1	209.3	10.2	15.6	12.7	472.7	117.6	55225	2212	21.7	4.34	2072	211.3
	82	528.3	208.7	9.6	13.2	12.7	472.7	104.3	47363	1826	21.3	4.18	1793	175.0
533 × 165	73	528.8	165.6	9.3	13.5	12.7	476.5	93.0	40414	1027	20.8	3.32	1528	124.1
	66	524.8	165.1	8.8	11.5	12.7	476.5	83.6	35083	863	20.5	3.21	1337	104.5
457 × 191	98	467.4	192.8	11.4	19.6	10.2	404.4	125.2	45653	2216	19.1	4.21	1954	229.9
	89	463.6	192.0	10.6	17.7	10.2	404.4	113.8	40956	1960	19.0	4.15	1767	204.2
	82	460.2	191.3	9.9	16.0	10.2	404.4	104.4	37039	1746	18.8	4.09	1610	182.6
	74	457.2	190.5	9.1	14.5	10.2	404.4	94.9	33324	1547	18.7	4.04	1458	162.4
	67	453.6	189.9	8.5	12.7	10.2	404.4	85.4	29337	1328	18.5	3.95	1293	139.9

Table F.25

UNIVERSAL COLUMNS
Parallel Flanges
DIMENSIONS AND PROPERTIES

Serial Size	Mass per metre	Depth of Section D	Width of Section B	Thickness Web t	Thickness Flange T	Root Radius r	Depth between Fillets d	Area of Section	Moment of Inertia Axis x–x	Moment of Inertia Axis y–y	Radius of Gyration Axis x–x	Radius of Gyration Axis y–y	Elastic Modulus Axis x–x	Elastic Modulus Axis y–y
mm	kg	mm	mm	mm	mm	mm	mm	cm²	cm⁴	cm⁴	cm	cm	cm³	cm³
356 × 406	634	474.7	424.1	47.6	77.0	15.2	290.1	808.1	275140	98211	18.5	11.0	11592	4632
	551	455.7	418.5	42.0	67.5	15.2	290.1	701.8	227023	82665	18.0	10.9	9964	3951
	467	436.6	412.4	35.9	58.0	15.2	290.1	595.5	183118	67905	17.5	10.7	8388	3293
	393	419.1	407.0	30.6	49.2	15.2	290.1	500.9	146765	55410	17.1	10.5	7004	2723
	340	406.4	403.0	26.5	42.9	15.2	290.1	432.7	122474	46816	16.8	10.4	6027	2324
	287	393.7	399.0	22.6	36.5	15.2	290.1	366.0	99994	38714	16.5	10.3	5080	1940
	235	381.0	395.0	18.5	30.2	15.2	290.1	299.8	79110	31008	16.2	10.2	4153	1570
Column Core	477	427.0	424.4	48.0	53.2	15.2	290.1	607.2	172391	68057	16.8	10.6	8075	3207
356 × 368	202	374.7	374.4	16.8	27.0	15.2	290.1	257.9	66307	23632	16.0	9.57	3540	1262
	177	368.3	372.1	14.5	23.8	15.2	290.1	225.7	57153	20470	15.9	9.52	3104	1100
	153	362.0	370.2	12.6	20.7	15.2	290.1	195.2	48525	17470	15.8	9.46	2681	943.8
	129	355.6	368.3	10.7	17.5	15.2	290.1	164.9	40246	14555	15.6	9.39	2264	790.4
305 × 305	283	365.3	321.8	26.9	44.1	15.2	246.6	360.4	78777	24545	14.8	8.25	4314	1525
	240	352.6	317.9	23.0	37.7	15.2	246.6	305.6	64177	20239	14.5	8.14	3641	1273
	198	339.9	314.1	19.2	31.4	15.2	246.6	252.3	50832	16230	14.2	8.02	2991	1034
	158	327.2	310.6	15.7	25.0	15.2	246.6	201.2	38740	12524	13.9	7.89	2368	806.3
	137	320.5	308.7	13.8	21.7	15.2	246.6	174.6	32838	10672	13.7	7.82	2049	691.4
	118	314.5	306.8	11.9	18.7	15.2	246.6	149.8	27601	9006	13.6	7.75	1755	587.0
	97	307.8	304.8	9.9	15.4	15.2	246.6	123.3	22202	7268	13.4	7.68	1442	476.9
254 × 254	167	289.1	264.5	19.2	31.7	12.7	200.2	212.4	29914	9796	11.9	6.79	2070	740.6
	132	276.4	261.0	15.6	25.1	12.7	200.2	167.7	22416	7444	11.6	6.66	1622	570.4
	107	266.7	258.3	13.0	20.5	12.7	200.2	136.6	17510	5901	11.3	6.57	1313	456.9
	89	260.4	255.9	10.5	17.3	12.7	200.2	114.0	14307	4849	11.2	6.52	1099	378.9
	73	254.0	254.0	8.6	14.2	12.7	200.2	92.9	11360	3873	11.1	6.46	894.5	305.0
203 × 203	86	222.3	208.8	13.0	20.5	10.2	160.8	110.1	9462	3119	9.27	5.32	851.5	298.7
	71	215.9	206.2	10.3	17.3	10.2	160.8	91.1	7647	2536	9.16	5.28	708.4	246.0
	60	209.6	205.2	9.3	14.2	10.2	160.8	75.8	6088	2041	8.96	5.19	581.1	199.0
	52	206.2	203.9	8.0	12.5	10.2	160.8	66.4	5263	1770	8.90	5.16	510.4	173.6
	46	203.2	203.2	7.3	11.0	10.2	160.8	58.8	4564	1539	8.81	5.11	449.2	151.5
152 × 152	37	161.8	154.4	8.1	11.5	7.6	123.4	47.4	2218	709	6.84	3.87	274.2	91.78
	30	157.5	152.9	6.6	9.4	7.6	123.4	38.2	1742	558	6.75	3.82	221.2	73.06
	23	152.4	152.4	6.1	6.8	7.6	123.4	29.8	1263	403	6.51	3.68	165.7	52.95

EQUAL ANGLES
DIMENSIONS AND PROPERTIES

Nominal Size	Area of Section	Centre of Gravity Cx	Centre of Gravity Cy	Actual Thickness	Mass per metre	Moment of Inertia Axis x-x	Moment of Inertia Axis y-y	Moment of Inertia Axis u-u Max.	Moment of Inertia Axis v-v Min.	Radius of Gyration Axis x-x	Radius of Gyration Axis y-y	Radius of Gyration Axis u-u Max.	Radius of Gyration Axis v-v Min.	Elastic Modulus Axis x-x	Elastic Modulus Axis y-y
mm	cm²	cm	cm	mm	kg	cm⁴	cm⁴	cm⁴	cm⁴	cm	cm	cm	cm	cm³	cm³
203 × 203	96.81	5.99	5.99	25.3	76.00	3686	3686	5845	1527	6.17	6.17	7.77	3.97	257	257
	91.09	5.93	5.93	23.7	71.51	3491	3491	5540	1442	6.19	6.19	7.80	3.98	243	243
	85.42	5.87	5.87	22.1	67.05	3294	3294	5232	1357	6.21	6.21	7.83	3.99	228	228
	79.69	5.81	5.81	20.5	62.56	3094	3094	4916	1271	6.23	6.23	7.85	3.99	213	213
	73.82	5.75	5.75	18.9	57.95	2885	2885	4587	1183	6.25	6.25	7.88	4.00	198	198
	67.89	5.69	5.69	17.3	53.30	2671	2671	4248	1093	6.27	6.27	7.91	4.01	183	183
	62.02	5.63	5.63	15.8	48.68	2455	2455	3907	1004	6.29	6.29	7.94	4.02	167	167
152 × 152	62.83	4.60	4.60	22.1	49.32	1321	1321	2089	553	4.58	4.58	5.77	2.97	124	124
	58.63	4.54	4.54	20.5	46.03	1243	1243	1968	517	4.60	4.60	5.79	2.97	116	116
	54.45	4.49	4.49	19.0	42.75	1164	1164	1846	482	4.62	4.62	5.82	2.98	108	108
	50.09	4.42	4.42	17.3	39.32	1080	1080	1714	446	4.64	4.64	5.85	2.98	99.8	99.8
	45.95	4.37	4.37	15.8	36.07	999	999	1587	411	4.66	4.66	5.88	2.99	91.9	91.9
	41.55	4.31	4.31	14.2	32.62	911	911	1448	374	4.68	4.68	5.90	3.00	83.3	83.3
	37.03	4.24	4.24	12.6	29.07	819	819	1303	335	4.70	4.70	5.93	3.01	74.5	74.5
	32.61	4.18	4.18	11.0	25.60	727	727	1156	297	4.72	4.72	5.96	3.02	65.7	65.7
	28.06	4.11	4.11	9.4	22.02	631	631	1003	258	4.74	4.74	5.98	3.03	56.7	56.7
127 × 127	44.80	3.85	3.85	19.0	35.16	651	651	1028	273	3.81	3.81	4.79	2.47	73.5	73.5
	41.37	3.79	3.79	17.4	32.47	607	607	961	253	3.83	3.83	4.82	2.47	68.1	68.1
	37.78	3.73	3.73	15.8	29.66	560	560	888	232	3.85	3.85	4.85	2.48	62.4	62.4
	34.14	3.67	3.67	14.2	26.80	511	511	811	211	3.87	3.87	4.87	2.48	56.6	56.6
	30.56	3.61	3.61	12.6	23.99	462	462	734	190	3.89	3.89	4.90	2.49	50.8	50.8
	26.93	3.55	3.55	11.0	21.14	411	411	654	169	3.91	3.91	4.93	2.50	44.9	44.9
	23.31	3.49	3.49	9.5	18.30	359	359	571	147	3.93	3.93	4.95	2.51	39.0	39.0
102 × 102	35.12	3.22	3.22	19.0	27.57	317	317	497	136	3.00	3.00	3.76	1.97	45.6	45.6
	32.45	3.16	3.16	17.4	25.48	296	296	466	126	3.02	3.02	3.79	1.97	42.3	42.3
	29.78	3.10	3.10	15.8	23.37	275	275	434	116	3.04	3.04	3.82	1.97	38.9	38.9
	26.96	3.04	3.04	14.2	21.17	252	252	399	105	3.06	3.06	3.84	1.97	35.4	35.4
	24.09	2.98	2.98	12.6	18.91	228	228	361	94.3	3.07	3.07	3.87	1.98	31.7	31.7
	21.27	2.92	2.92	11.0	16.69	203	203	323	83.8	3.09	3.09	3.90	1.99	28.1	28.1
	18.39	2.86	2.86	9.4	14.44	178	178	283	73.1	3.11	3.11	3.92	1.99	24.4	24.4
	15.37	2.79	2.79	7.8	12.06	150	150	239	61.7	3.13	3.13	3.95	2.00	20.4	20.4

Table F.26

CHANNELS
DIMENSIONS AND PROPERTIES

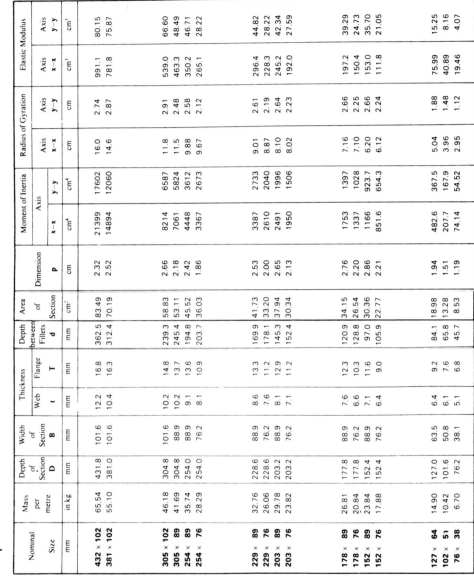

Nominal Size	Mass per metre	Depth of Section D	Width of Section B	Thickness Web t	Thickness Flange T	Depth between Fillets d	Area of Section	Dimension p	Moment of Inertia Axis x–x	Moment of Inertia Axis y–y	Radius of Gyration Axis x–x	Radius of Gyration Axis y–y	Elastic Modulus Axis x–x	Elastic Modulus Axis y–y
mm	in kg	mm	mm	mm	mm	mm	cm²	cm	cm⁴	cm⁴	cm	cm	cm³	cm³
432 × 102	65.54	431.8	101.6	12.2	16.8	362.5	83.49	2.32	21399	17602	16.0	2.74	991.1	80.15
381 × 102	55.10	381.0	101.6	10.4	16.3	312.4	70.19	2.52	14894	12060	14.6	2.87	781.8	75.87
305 × 102	46.18	304.8	101.6	10.2	14.8	239.3	58.83	2.66	8214	6587	11.8	2.91	539.0	66.60
305 × 89	41.69	304.8	88.9	10.2	13.7	245.4	53.11	2.18	7061	5824	11.5	2.48	463.3	48.49
254 × 89	35.74	254.0	88.9	9.1	13.6	194.8	45.52	2.42	4448	3612	9.88	2.58	350.2	46.71
254 × 76	28.29	254.0	76.2	8.1	10.9	203.7	36.03	1.86	3367	2673	9.67	2.12	265.1	28.22
229 × 89	32.76	228.6	88.9	8.6	13.3	169.9	41.73	2.53	3387	2733	9.01	2.61	296.4	44.82
229 × 76	26.06	228.6	76.2	7.6	11.2	178.1	33.20	2.00	2610	2040	8.87	2.19	228.3	28.22
203 × 89	29.78	203.2	88.9	8.1	12.9	145.3	37.94	2.65	2491	1996	8.10	2.64	245.2	42.34
203 × 76	23.82	203.2	76.2	7.1	11.2	152.4	30.34	2.13	1950	1506	8.02	2.23	192.0	27.59
178 × 89	26.81	177.8	88.9	7.6	12.3	120.9	34.15	2.76	1753	1397	7.16	2.66	197.2	39.29
178 × 76	20.84	177.8	76.2	6.6	10.3	128.8	26.54	2.20	1337	1028	7.10	2.25	150.4	24.73
152 × 89	23.84	152.4	88.9	7.1	11.6	97.0	30.36	2.86	1166	923.7	6.20	2.66	153.0	35.70
152 × 76	17.88	152.4	76.2	6.4	9.0	105.9	22.77	2.21	851.6	654.3	6.12	2.24	111.8	21.05
127 × 64	14.90	127.0	63.5	6.4	9.2	84.1	18.98	1.94	482.6	367.5	5.04	1.88	75.99	15.25
102 × 51	10.42	101.6	50.8	6.1	7.6	65.8	13.28	1.51	207.7	167.9	3.96	1.48	40.89	8.16
76 × 38	6.70	76.2	38.1	5.1	6.8	45.7	8.53	1.19	74.14	54.52	2.95	1.12	19.46	4.07

Table F.27

Table F.28

CHANNELS
BUCKLING VALUES FOR UNSTIFFENED WEBS

FOR GRADE 43 STEEL

Nominal size	Mass per metre	Web thickness t	Depth between fillets d	End bearing			Continuous over bearing		
				Channel component C_1	Stiff bearing component C_2	Flange plate component C_3	Channel component C_1	Stiff bearing component C_2	Flange plate component C_3
mm	kg	mm	mm	kN	kN/mm	kN/mm	kN	kN/mm	kN/mm
432 × 102	65.54	12.2	362.5	348	1.61	1.61	696	1.61	3.23
381 × 102	55.10	10.4	312.4	261	1.37	1.37	522	1.37	2.74
305 × 102	46.18	10.2	239.3	215	1.41	1.41	430	1.41	2.83
305 × 89	41.69	10.2	245.4	215	1.41	1.41	430	1.41	2.82
254 × 89	35.74	9.1	194.8	162	1.28	1.28	324	1.28	2.55
254 × 76	28.29	8.1	203.7	141	1.11	1.11	282	1.11	2.22
229 × 89	32.76	8.6	169.9	139	1.22	1.22	278	1.22	2.43
229 × 76	26.06	7.6	178.1	120	1.05	1.05	240	1.05	2.11
203 × 89	29.78	8.1	145.3	117	1.15	1.15	234	1.15	2.31
203 × 76	23.82	7.1	152.4	101	1.00	1.00	202	1.00	1.99
178 × 89	26.81	7.6	120.9	97	1.09	1.09	194	1.09	2.18
178 × 76	20.84	6.6	128.8	83	0.93	0.93	166	0.93	1.87
152 × 89	23.84	7.1	97.0	79	1.03	1.03	158	1.03	2.06
152 × 76	17.88	6.4	105.9	70	0.92	0.92	140	0.92	1.83
127 × 64	14.90	6.4	84.1	59	0.93	0.93	118	0.93	1.87
102 × 51	10.42	6.1	65.8	46	0.90	0.90	92	0.90	1.80
76 × 38	6.70	5.1	45.7	29	0.76	0.76	58	0.76	1.52

If L_b is length (mm) of stiff bearing and t_p is total thickness (mm) of flange plates and packing between beam and bearing, then:

Web capacity = $C_1 + L_b C_2 + t_p C_3$.

The web buckling and direct bearing values are applicable also to other points of concentrated load.

For full explanation of buckling and direct bearing values see 'Metric Practice for Structural Steelwork: 2nd Edition' (BCSA).

Table F.28 (*contd.*)

CHANNELS
DIRECT BEARING VALUES

FOR GRADE 43 STEEL

Nominal size	Mass per metre	End bearing			Continuous over bearing			Shear value
		Channel component C_1	Stiff bearing component C_2	Flange plate component C_3	Channel component C_1	Stiff bearing component C_2	Flange plate component C_3	
mm	kg	kN	kN/mm	kN/mm	kN	kN/mm	kN/mm	kN
432 × 102	65.54	139	2.32	4.01	278	2.32	8.03	526
381 × 102	55.10	117	1.98	3.42	234	1.98	6.85	396
305 × 102	46.18	110	1.94	3.36	220	1.94	6.71	310
305 × 89	41.69	100	1.94	3.36	200	1.94	6 71	310
254 × 89	35 74	89	1.73	2 99	178	1.73	5.99	231
254 × 76	28.29	67	1.54	2.67	134	1.54	5.33	205
229 × 89	32.76	83	1.63	2.83	166	1.63	5.66	196
229 × 76	26.06	63	1.44	2.50	126	1.44	5.00	173
203 × 89	29.78	77	1.54	2.67	154	1.54	5.33	164
203 × 76	23.82	59	1.35	2.34	118	1.35	4.67	144
178 × 89	26.81	71	1.44	2.50	142	1.44	5.00	135
178 × 76	20.84	53	1.25	2.17	106	1.25	4.34	117
152 × 89	23.84	65	1.35	2.34	130	1.35	4.67	108
152 × 76	17.88	49	1.22	2.11	98	1.22	4.21	97
127 × 64	14.90	45	1.22	2.11	90	1.22	4.21	81
102 × 51	10.42	36	1.16	2.01	72	1.16	4.01	61
76 × 38	6.70	26	0.97	1.68	52	0.97	3.36	38

If L_b is length (mm) of stiff bearing and t_p is total thickness (mm) of flange plates and packing between beam and bearing, then:

Web capacity = $C_1 + L_b C_2 + t_p C_3$.

The web buckling and direct bearing values are applicable also to other points of concentrated load.

For full explanation of buckling and direct bearing values see 'Metric Practice for Structural Steelwork: 2nd Edition' (BCSA).

Table F.29

JOISTS

DIMENSIONS AND PROPERTIES

Nominal Size	Mass per metre	Depth of Section D	Width of Section B	Thickness		Depth between Fillets d	Area of Section	Moment of Inertia		Radius of Gyration		Elastic Modulus	
				Web t	Flange T			Axis		Axis x–x	Axis y–y	Axis x–x	Axis y–y
								x–x	y–y				
mm	kg	mm	mm	mm	mm	mm	cm²	cm⁴	cm⁴	cm	cm	cm³	cm³
203 × 102	25.33	203.2	101.6	5.8	10.4	161.0	32.3	2294	2023	8.43	2.25	225.8	32.02
178 × 102	21.54	177.8	101.6	5.3	9.0	138.2	27.4	1519	1340	7.44	2.25	170.9	27.41
152 × 39	17.09	152.4	88.9	4.9	8.3	117.9	21.8	881.1	762.1	6.36	1.99	115.6	19.34
127 × 76	13.36	127.0	76.2	4.5	7.6	94.2	17.0	475.9	399.8	5.29	1.72	74.94	13.17
102 × 64	9.65	101.6	63.5	4.1	6.6	73.2	12.3	217.6	181.9	4.21	1.43	42.84	7.97
76 × 51	6.67	76.2	50.8	3.8	5.6	50.3	8.49	82.58	68.85	3.12	1.14	21.67	4.37

Appendix G
Definitions

access door: removable section of formwork to give access for cleaning out, placing concrete or inserting a vibrator.

anchor: fixing device made by embedding nut with loop, or other similar connection, in concrete.

base plate: metal plate with spigot for distributing load from scaffold tube, may be adjustable.

bay: section between construction joints in a concrete wall or slab.

blinding concrete: sealing layer of concrete placed on excavated ground to provide a clean and accurate working surface for reinforcement fixing.

bolt (shutter): see *tie.*

box-out: formation of aperture or recess in concrete, usually to permit later building-in of component.

brace: diagonal member used to triangulate compression elements and provide frame stability.

bridle: short beam spanning between main beams to support intermediate beams which have been cut to make an opening.

camber: amount by which soffit formwork is set higher than the finished profile to allow for take-up in supports and deflection of permanent work under load.

cantilever formwork: single-sided wall formwork supported by projecting cantilever soldier tied in to previous lift.

centering: support for deck or floor formwork. Particularly applicable to arches and other curved soffits.

check: small timber section fixed to formwork at the top of a lift to form a sharp joint between lifts.

climbing formwork: self-supporting formwork used for the construction in successive multiple lifts of single- or double-sided face work.

cone: part of a formwork tie assembly, used to connect re-usable and embedded parts.

construction joint: internal face in concrete between adjacent bays either to control shrinkage or divide the work into convenient sections, designed to act monolithically.

coupler: component used to joint scaffold tube, may be plain – tubes at right angles; swivel – tubes angled; parallel – tubes side by side.

crack inducing joint: the concentration of shrinkage in long walls at a predetermined point by the formation of a vertical chase on the face.

decking: sheeting to soffit formwork.

diagonals: diagonal bracing to scaffold assembly.

drop-head: device fitted to a tubular prop to permit the removal of beams and soffit formwork without disturbing the prop.

dumb bolt: inactive bolt used to set anchor in position.

eccentricity: amount by which load is applied off-centre to a prop.

factor of safety: the ratio ultimate load : design working load.

falsework: any temporary structure used to support a permanent structure while it is not self-supporting.

fix and strike: traditional description for erection and removal of formwork. Used in bonus task terminology and some methods of measurement.

floor centres: telescopic beam used as support to soffit decking.

flying forms: large table form designed to be extracted and re-erected without dismantling by moving with a crane.

forkhead: U-shaped fitting to support joists on top of a tube or prop, usually adjustable.

formwork: that part of the temporary work for *in situ* or precast concrete which forms the finished shape. It includes the sheathing or decking and the members providing direct support.

frame: two or more principal members in the same plane braced to form a structural panel.

grade stress: the stress that can be safely sustained by timber of a particular strength class, or species and grade.

imposed load: any load that a member is required to resist in addition to its own self-weight.

jack (cantilever): screw fitted to the lower end of soldiers to adjust vertical alignment.

joist: any steel, timber, or aluminium section used as a beam to support vertical loads.

kentledge: weight to give counterbalance or provide a reaction.

key: recess formed in the face of a construction joint to improve shear.

kicker: upstand formed on foundation to locate and fix subsequent wall or column construction.

lacing: horizontal scaffold tube used to tie vertical load-bearing standards (or other supports) into a lattice and reduce the effective length.

lift: the height of concrete placed in one operation.

make-up: formwork made on the spot to suit as-found dimensions and irregular shapes.

movement joints: gap formed between concrete sections to accommodate movement due to shrinkage or thermal expansion/contraction.

permissible stress: the stress that can be sustained with acceptable safety by a structural component under the particular condition of service or loading.

permit: formal authorisation to load or strike falsework.

prop: any vertical or inclined compressive support, but usually proprietary adjustable tubular strut.

raking shore: inclined strut or prop.
runner: longitudinal timber continuous over several supports.

sheathing: surface material of formwork in contact with the concrete.
sheeting: as *sheathing.*
shuttering: as *formwork* but may include main members.
slipform: formwork assembly designed to extrude concrete section during continuous movement.
snap tie: a formwork tie designed to be broken off after use.
soffit: the exposed under-surface of any concrete element.
soldier: principal structural component of wall formwork. It acts as a vertical beam or cantilever to transmit concrete pressure on to ties or anchors.
standard: a vertical or near-vertical tube.
stop end: vertical formwork to form a temporary or hidden face in concrete.
strength class: grading category for timber, determines permissible stresses.
stringer: sloping edge form in staircase construction.
strong-back: heavy duty soldier reinforced for lifting.
strut: a compression member.
stud: intermediate stiffening member in a framed formwork panel.

table form: combined soffit and support falsework, used in modular multi-storey construction.
tie: rod or bar component set between formwork faces to resist concrete pressure.

waling: continuous horizontal beam member to formwork, transmits concrete pressure on to ties.

yoke: continuous frame around column formwork, acts as combined waling and ties.

Index

abutments 98, 99
acceleration 95, 96, 98
adjustable floor centres 53
adjustable steel props 58
aggregate transfer 8
aluminium 7, 33
analysis 104
anchors 28
angles 167
approximate analysis 140, 142
arch girders 63
arching 38
axial force, axial stress 104, 105, 106, 107,
    108, 142, 151

base 111, 112, 113, 114
beam 105, 109, 111, 119, 120, 121, 156, 159, 160
beam and slab floors 52
beam clamps and struts 53
bearing, bearing capacity 109, 115, 149, 153,
    158, 159, 160
bending 91, 98, 103, 104, 105, 107, 109, 112,
    116, 117, 120, 152
birch-faced plywood 132
blasting 91
bolts 103, 104, 109, 110, 111, 112, 113, 119,
    121, 150, 153
bolts, tie, design 41
boom 104, 106, 107, 110, 112, 141
box, box section 114, 144
bracing, bracing diagonal 69, 103, 104, 105,
    106, 107, 109, 110, 111, 112, 113,
    115, 116, 120, 141
Bragg Committee 84
bridges 90, 95, 97, 98, 99, 102, 109, 110, 112,
    114, 115
bridging girders 63
brief, design 87
British Standards 94
buckling 94, 108, 111, 112, 118, 119, 120, 144,
    159
buoyancy 95
bush hammering 8

cables 117
camber 71
Canada 94
cantilever 111, 112, 117
cap 107, 108, 109, 111
centering 1
centering materials 56
centroid 119
channels 168, 169
checking 121
clay 116
cliffs 101
coastal 95
cofiform 134
collapse 97, 120
collision 91
column 94, 96, 98, 113
    box 25
    clamps 24
    shutters 23
compression, compressive stress 105, 107, 108,
    109, 110, 111, 112, 113, 115, 116, 120,
    140, 141, 151, 153, 158
computer 112
concrete 90, 91, 92
    finish 6
    mass 150
    pressures 37
connections 109, 121
constraint 98
construction 99, 109
construction plant 91
continuity 74
contraction 109
couplers 105, 113, 115, 155
cracking 91, 117, 118, 121
crane 117, 120
crash barriers 118, 148
creep 97
cross bracing 106
cross fall 103
cross-sectional area 105, 106, 120, 142
crushing 108, 111, 112, 118, 119, 120
current 95

curvature 112
cyclic forces 92

damage, damage tolerances 92, 96, 97, 104, 118
damping 94
data tables 148
dead loading 91, 121
debris 95
deck 98, 99, 101, 102, 103, 104, 107, 109, 112, 113, 117, 120
deck falsework 52
decking systems 56
definitions 172
deflections 90, 91, 98, 103, 104, 105, 112, 113, 119, 120, 121, 142, 158
depth to breadth ratios 4
design life 95, 99, 101
design methods, design philosophy 90, 105
diagonal bracing 69
diaphragm walls 120
dimension 95
dumb bolt 28
dynamic, dynamic response 92, 94, 101, 148

early strip 56
earth pressure 92
earthquake 94, 95, 96
eccentricity 115, 120
economic 105, 118
effective – depth, length, area 104, 105, 108, 109, 110, 113, 120, 153, 155, 157, 158, 159
elastic analysis, elastic design 90, 98, 105
embankment 101
engineer 87
environmental 92, 99
equivalent strength class 3
erection tolerances 71, 138
expanded metal 53
expansion 109, 149
exposed aggregate 8

factor of safety 103, 111
failure 92, 118, 121
falsework (see also bridges)
    coordinator 86
    designer 86
    loading 73
    supervisor 86
    supplies 87
fatigue 92, 118
fillet 107, 108, 143
finite element 141
fire 92
fixity 104, 119
flange 105, 107, 108, 113, 120, 144, 157, 158
flexibility 112, 113, 142
flexural 106

flood 92, 95
floor slabs 52
footings 92, 98, 109, 116
force actions 90, 104
force coefficient 101
forkheads 117, 138
form liners 8
formwork
    columns 23
    permanent 9
    systems 35
    ties 29
    wall 26
foundations 70, 111, 112, 113, 118, 119, 120, 121
frame 104, 140, 142, 147
frictional resistance, frictional coefficients 95, 101, 149
frost heave 92
fundamental 95
funnel 99

gantry 117
geometry 94, 95, 98
girders 102, 103, 104, 107, 109, 112, 113, 117, 144
glass reinforced plastic 7
grade stresses 2
grillage 117, 119
ground conditions 92
ground roughness 92, 94
ground water 92
grout 117
gussets 103, 110
gust 99
guys 118

hangers 53, 119, 120
heaping 91, 99, 148
height limit 38
hinged 114
holes 106, 150
hollow sections 144, 160–163
horizontal forces 74, 101, 102, 103

ice, icing 95, 99, 102, 121, 148
immersed 95
impact 91, 95
imposed loading 91, 95, 99
independent check 88
inertia 110, 112
initial camber 71
injury 121
insurance 94
intensity 96
interfaces 95

jack, jacking 117, 119
job specifications 86

joists 171

kicker 23

lacing design 67
ladder 119
lateral stability 75
lattices 95, 101, 102, 103, 104, 113, 114, 142,
    147
legs 107, 109, 110, 116, 117
loading, falsework 71
loadings 91, 99, 109, 118, 148
load sharing 3
load/span graphs 125
lock nuts 119
loop anchor 28

magnitude 96
main loadings 91
mass 94
material properties 105
matrix 111
Mercalli 96, 98
method of sections 106, 109, 141
modulus of elasticity, timber 2, 3
moments 104, 107, 115

natural frequencies 92
neutral axes 103, 119, 120
node 103, 106, 115, 119, 120, 121
non-collapse design 94
Norway 102

optical illusion 112
optical instruments 120
oscillations 92
out of plumb 98, 111
overload 90, 91
overpour 91
overstress 91, 117, 121
overturning 94, 112, 117, 119, 121

packing 159, 160
period 95
permits 89
piers 98, 99, 100, 102, 103, 109, 111, 112,
    113, 114, 115
pile driving 91
piles 116, 120
pin-jointed 104, 112
plastic analysis 90, 105, 109
plate girder 102, 144
plumb 117, 118, 120
plywood
    Douglas fir 5
    Finnish 5
    re-use 7
    safe load table 132, 134
    sheathing design 46

pole 115, 116
portal 115, 116
post-tensioning 119, 120
Pratt truss 102
precamber 103, 112
prestressed 117
probability 94, 99, 121, 147
probable 92, 109, 118
propping 121
props, safe loads 138
pumping 92, 148

radius of gyration 105, 108, 110, 111, 152
records 95
refreezing 102
reinforcement 99
resonance 92
responsibility chart 85
return period 94
Richter 96, 98
river 95, 97
road 112
rocker 115
Rossi-Forel 96
rotation 98, 158
rubber 109
rust 118

safety 106, 112, 118, 119, 121
scaffold
    birdcage 64
    design example 64
    poles, tubes 99, 103, 105, 110, 113, 152,
        153
    proprietary systems 60
scouring 95
secondary beams 55, 73
second moment of area 108, 111
section 104
seismic, seismic risk 94, 95, 96, 98
serviceability 94, 98
settlement 98
shear centre 119, 120
shear forces, shear stress 95, 103, 104, 106,
        109, 112, 120, 143, 144
shear resistance 95
sheathing design 41
shielding factor 101, 145
shores 62
shrinkage 98
skip 100, 119, 120, 121
slam 95
slender, slenderness 104, 108, 153
slenderness ratio 113, 159
sliding 112
slip form 8
slump 38
snap ties 31
snow loading 95, 101, 102, 121, 148

soffit
    beams 59
    formwork 73
soldiers
    design 45
    proprietary 31
spacers 109
span 99, 102, 107, 110
spring washers 119
stability 70, 98, 103, 104, 106, 109, 110, 115,
        116, 119, 120, 121, 144
standard solutions 76
statistical 101
steel properties, steel data 150, 151, 154
stiffeners 103, 107, 108, 109, 110, 112, 116,
        119, 120
stiffening limit 38
stiffness 94, 120
stiffness analysis 112
storage of materials 91
strain 117
stress grading 4
strut 105, 111, 120, 151
surface finishes 7
sway 113
symmetry 106

telescopic 113
temperature 98, 109, 113
temporary 99
tension 105, 106, 110, 117, 121, 140, 141
tests 94
tie 29, 103, 113, 115, 120, 157
timber
    defects 2
    falsework stresses 3
    geometric properties 136
    grade stresses 2
    load/span graphs 125
    modulus of elasticity 3
    moisture content 2
    strength properties 2
    stress grading 2
timbering 1
torsion 119, 120, 143, 144
towers 62, 103, 107, 108, 109, 110, 111, 113,
        114, 117, 119

trenches 120
trestles 63
trough and waffle floors 53
trusses 99, 102, 103, 120, 141
tubes 102, 103, 106, 107, 110, 111, 113, 115,
        116, 144, 152, 153
twist 120, 144

Universal sections 99, 105, 116, 147, 164–166
uplift 94, 112

variations 91
velocity 95
vertical loading, force, deflection 99, 101, 102,
        104, 105, 106, 107, 109, 112
vibration 91, 94, 96, 119
vortex 95

walings 28
    design 41
walkways 101
wall 120, 144
    formwork 26
    formwork solutions 40, 41
    panels 27
    proprietary systems 35
    ties 29
warping 143
Warren girders 102, 105, 106, 107, 108, 109,
        112, 114
water 95
water bar tie 30
wave 95, 97
web 106, 107, 108, 111, 112, 116, 119, 120,
        159, 160
weight 91, 95, 99
welds 103, 104, 110, 113, 118, 121
wheels 119
wind forces 92, 99, 112, 113, 145, 147
wind loads 67, 94, 101, 102, 106, 110, 113,
        118, 119, 121, 146
wind speed 92, 93, 94, 101, 147
winter 95
workability 37
workmanship 138

yielding 94